Meteorites and their Origins

by

GJH McCall DSc

*Formerly Reader in Geology
at the University of Western Australia*

David & Charles: Newton Abbot

ISBN 0 7153 5560 0

Set in 10 on 12 point Times
and printed in Great Britain
at The Pitman Press Bath
for David & Charles (Holdings) Limited
South Devon House Newton Abbot, Devon

Meteorites and their Origins

Contents

Illustrations

9

ILLUSTRATIONS

1: Introduction

The science of meteoritics is shared territory, concerning which astronomy, chemistry, geology, and physics all have their spheres of interest. Meteoritics is an aspect of science that fascinates the layman—the amateur geologist or astronomer—and likewise the professional scientist, whether he is of the disciplines closely involved or other less involved disciplines such as the life sciences. Meteoritics is not usually taught in formal lecture courses at Tertiary (University) level, yet many scientists require a general knowledge of the subject.

Some geologists can recognise a meteorite, most not: which is a pity, for it is to the geologist that possible meteorite material is likely to be brought in for examination in the first instance. One cannot go out and search blindly for meteorites in the same way as one approaches geological field studies: one must start with the initial lead or clue, which comes either from a fall being reported or a lump of strange, 'foreign looking' material being found and brought in, in whole or part, for examination. The meteoriticist must be prepared to endure many 'wild-goose chases' and follow up many 'red herrings' to obtain even one success. Chemists and physicists are commonly versed in the geochemical and physical aspects of meteoritics, but a surprisingly small percentage have a really good background knowledge of the general aspects of the subject. This lack tends to vitiate otherwise admirable specialised treatments of the subject which they publish, the final discussions of broader implications tending to naïvity. Astronomers are not so much interested in the meteorite after it falls from the sky as with the mechanics of its arrival, its orbital characteristics and the correlation between its chemistry and their indirect observations of the spectrochemistry of comets, meteors, etc. Especially too, are they interested in the traces left on the mass by the meteorite's fierce rush through the Earth's atmosphere.

The author has endeavoured, in writing this book, to condense the

13

vast spectrum of meteoritics into a single readable volume beyond the mere elementary statement, but still suitable for use as a general text for amateur scientists, university students, and professional scientists. He has attempted to limit the use of jargon and complex terminology, but meteoritics, like geology, is bedevilled with strange names. Detailed references have been avoided, but a bibliography is added at the end.

A book of this nature is not original material, being derived from a wealth of scientific papers and previously published texts. The character of this book is somewhat different from earlier treatises, in that virtually all aspects of interest to the general reader are at least touched on. These include historical, astronomical, and physical aspects, the latter including the physical nature of the masses; the manner of fall, petrography, mineralogy, chemistry, classification, and physical properties of meteorites are covered; more specialised sections of the text cover theories of origin and planetological considerations. Life in meteorites and age-dating results are covered in general terms. There have been a number of excellent texts published on this subject and the author has drawn heavily on these: among those that must be singled out are the following; the numbers refer to the general reading list in the bibliography at the end of this book.

Mason (1)   — essentially petrological and geochemical.
Krinov (2)   — concerned primarily with the physical aspects of the masses.
Fedynsky (3) — concerned primarily with astronomical aspects.
Nininger (4) — reflecting the author's evangelical enthusiasm and flair for the search.
Heide (5)   — a brief but admirable elementary treatment.

This text has no special emphasis, even though the author approaches the subject as a petrologist and volcanologist, having admittedly to step beyond his own special field to produce a balanced and comprehensive account: however, no writer of a broad account of this subject can avoid such a step.

The author acknowledges his debt to the authors named before, and to the authors of countless scientific works of lesser bulk but no less worth: in particular, he has drawn heavily on review papers by Anders, Wood, and Krinov (6, 7, 8, and 9) concerning true meteorites: on papers by Perry and Axon (10, 11) concerning their metallurgy: by

Ringwood and Mason concerning their origins (12, 13, 14): and by Baker and O'Keefe (15, 16) concerning tektites, which are covered in a special chapter, despite some reserve concerning their meteoritic status. Countless other scientific papers have been utilised, and as many as possible are listed in the bibliography.

This then is a post-Apollo 11 review of meteoritics.

2: Historical—Stones from the Sky

Stones that 'fell from the sky' provided objects of veneration for many early dwellers on this Earth, but it took a surprising number of centuries of civilisation for the reality of such visitations to attain universal acceptance by scientists. Our surprise at this scepticism must be tempered by the realisation that fact and fable, myth and historical record were so intimately entwined, even by such advanced peoples as the Greeks and Romans, that later historians and scientists found it all too easy to dismiss reports of extraordinary cosmic visitations as exaggeration or illusion. Even after Newtonian principles became generally accepted, in the late eighteenth and early nineteenth centuries, there was some delay in accepting the reality of these phenomena—but no more delay, perhaps, than the propounder of any novel and original hypothesis must expect, even in this day and age! That scepticism has been the rule for so long does not surprise the author: in dealing with reports of supposed observations by people of this, the twentieth, century, it becomes evident that even the most respected members of the community can indulge, on occasion, in embroidery of a tale and even in complete romancing, and besides this, constant repetition of a tale passing from one person to another converts mere rumour to gospel truth! Though scientists may appear to have been unduly sceptical there does seem to be considerable justification for this scepticism!

Of *meteors*—shooting stars, bright objects seen to dart across the night sky—records go back to the Egyptian papyrus writings of *c* 2000 BC (4). Records of actual masses of stone or iron, that is meteorites, falling to the Earth from out of the sky go back almost as far. Of the latter records, that of 'a black stone in the form of a cone, circular below and ending in an apex above', in Phrygia about 652 BC brings us, out of the long distant past, the very well-drawn image of an oriented meteorite. Livy mentions falls of stones about the same time

17

on the Alban Mount, in Italy, and one would like to believe that it was of meteoritic origin, but Fletcher (17) in 1904 reasonably questioned this interpretation, suspecting the event to have been in the nature of a shower of lapilli of volcanic origin. Whatever its origin, this shower so impressed the senate that it declared a public holiday for nine days (the original 'nine days wonder'?). Plutarch and Pliny both mention falls of stones, and the latter refers to one as being 'of the size of a wagon': where did it go to? It would be the largest known stone meteorite!

Even earlier, though less definite, records come to us from out of the Parian chronicle—of falls in Crete about 1478 BC, and again in Boeotia about 1200 BC, and once again in Crete in 1168 BC. The latter was an iron meteorite: it is interesting to note that falls of stones seem to predominate in the ancient records just as at the present time. In 616 BC a meteoritic stone is recorded as having broken several chariots and killed ten men. This appears to be the only record of a meteoritic fatality: in recent times the author can only find a report of a lady sustaining a broken arm!

The sacred stone of the Kaaba, in Mecca, Arabia, is reputed to have fallen out of the sky. It was apparently long known prior to the birth of Islam. There is, however, no certainty that it is of meteoritic character, forbidden as it is to infidel eyes by the force of strict religious interdict.

Falls in China are particularly well documented from the time of the Han Dynasty (89 BC–AD 29) according to Nininger, who gives a list that is surprisingly long for such a brief span of time. This prolific record is also in striking contrast to the extraordinary blank in the meteorite distribution record afforded by China at the present day (see statistics of fall and find distribution).

There is little need to dwell on the religious status afforded to meteorites, for this is common knowledge. From the days of the ancient Greeks, at least, right up to the present time, Man has reacted to these visitations by enshrining the ominous visitor. It is reported from India that it was necessary for officers of the Geological Survey to go hot-foot to the spot, immediately a report of a fall was received, for otherwise the mass would either have been enshrined or broken to small pieces to release the evil spirits which it contained. Either way, it was lost to science. The American Indians made a practice of transporting meteorites long distances, and so, unwittingly, provided scientists with many meaningless locations of meteorite finds, some of

18

which were later shown to be in burial cists. Enshrinement is also reported from Africa. The Australian aborigines are known to carry tektites on their persons in their endless journeyings, believing that these mysterious objects give the possessor special magical powers. The earliest fall for which material is still preserved is believed to be that at Ensisheim in Alsace, during the year AD 1462. Since that time the actual material recovered has been stored in the local church. Particularly pleasing is the custom reported from medieval France, of chaining meteorites up—either to prevent them departing as swiftly as they came or else to prevent them wandering at night!

Iron meteorites are believed to have provided a source of weaponry and armoury in ancient times and this usage, coupled with centuries of sway by war lords, may well have a bearing on the geographic gaps in fall and find distribution, in areas such as China. Yet it does not seem to offer an entirely adequate resolution of the anomaly.

Records of meteor showers were kept by the ancient Russian and Chinese peoples, and the renowned Leonid shower can, in the Russian records, be traced back as far as the year AD 1202, when the display was spectacular. The Leonids kept their tryst punctually every thirty-three years up to AD 1866, but have since only been observed in their full glory by observers in restricted areas—for example observations in 1966, in North America.

Among many other graphic descriptions in the early Russian records, recorded by Krinov, are the account of a fall at Vyshegorod in the eleventh century and the fall of a giant meteorite near Velikii Ustiug in 1296.

> On the second week, at noon, there appeared over the town—a dark cloud, and it was dark as night: after this there appeared great clouds rising on all four sides and from these clouds lightning kept ceaselessly flashing. As it thundered over the town it was impossible to talk. Even the ground seemed to shake and sway continuously as if terrified by this horror. And clouds of fire arose and collided with one another: great heat arose from the lightning and thunder.

This event was recorded as the fall of 'a stony swarm' in the forests, near the town. It is not unlike descriptions of the twentieth-century Tungus and Sikhot-Alin events, also in Siberia.

Fig 1 *Ermak Timofeevich inspecting the meteorite which fell near the town of Tashatkan in* AD *1584. The drawing is from the annals of Semen Remezov (from E. L. Krinov, 1960)*

A drawing, reproduced here after Krinov, relates to the fall of a stone, 'purple in colour and as big as a sleigh', that fell in AD 1584 near the town of Tashatkan (which name, derived from this event, signifies a stony arrow). The description of a fall at Novye Ergi in AD 1662 gives us, when imagery is detached from fact, a picture of a small cloud, 'from which fell stones, shining bright and light, large and small, all hot'. What could better describe to us how meteorites are now known to fall, by day?

Scientific Acceptance

Acknowledgement of the reality of falls of stones from the sky came only after publications by Chladni and Biot. The former made his pronouncements in a book published in Riga in the year AD 1794. Chladni was of German stock, but was a corresponding member of the Academy of Science at St Petersburg (now Leningrad), and Russia, justifiably, claims him. He does not seem to have been the first person to recognise meteorites for what they really are, but his pioneer contribution did lead, at last, to a real understanding of these phenomena. His statement was largely based on consideration of two large masses—the type Pallasite, the original 680 kg 'Pallas' iron from Krasnojarsk Siberia, and another immense find of iron material, estimated at 13,500 kg weight, at Otumpa in the Gran Chaco province of South America. Chladni cited evidence of slow cooling, of even distribution of olivine throughout the Pallas meteorite, and the fact that the Otumpa iron was completely exotic, a 'foreigner' in schist terrain devoid of any trace of terrestrial iron mineralisation. Rejecting the work of Man, electricity, and accidental conflagration, he concluded that both masses must have been projected to their site of recovery from a far distant source. He believed that no volcanoes eject iron in native form (pure iron containing minor amounts of nickel): however, a later writer, Fletcher, early in the present century, recorded iron as being ejected from Monte Somma, Italy and we know of terrestrial, and presumably volcanic, nickel-iron from basalts in Disko Island, Greenland (from the writings of Bøggyld). Nevertheless, the fact that there is a complete lack of volcanoes in either region supports the rejection of volcanic ejection. Chladni was thus left, by a process of elimination, with a single possible answer—a cosmic source in the sky. He suggested that the well-known phenomenon of fireball meteors had some connection with these objects, and we know that this was the correct answer.

21

Though Chladni is commonly afforded the honour of establishing the reality and nature of meteorite falls, we must, with due respect to his masterly analysis of the critical evidence, acknowledge that he had several no less discerning predecessors. Troili had already published an account of a fall at Albareto, in Italy, and even provided a mineralogical description of the stone recovered. Yet this factual and realistic description met with scant credence. An even earlier writer, de la Lande, reported on a fall at Bresse, in France, and, centuries earlier, the philosopher Diogenes of Apollonia had really solved the problem! He recognised the cosmic origin of meteorites and meteors— 'meteors are invisible stars that fall to Earth and die out, like the fiery, stony star that fell to Earth near the Egos Potamos river.'

Troili and Chladni were followed by a period of scepticism. A lucid account of the fall of a stone at Lucé, in France, by Father Bachelay, even though supported by the actual stone (and the respect due to his cloth), failed to convince certain sceptical scientists. Lavoisier and his associates analysed the stone and described it as 'an ordinary one, struck and altered by lightning, and showing nothing unusual on analysis'. Again and again we find, in such writings, the confusion between thunder and lightning, traditionally related to solid 'bolts', and meteorites. The very word *meteorite*, so similar to the word *meteorology*, shares the same root, on account of this concept of the solid 'bolt'.

As if to underline the statement by Chladni, received with limited credence, Nature entered the controversy! A fall of stones occurred in Tuscany in 1794, the year of Chladni's publication; strangely, according to the Earl of Bristol, only eighteen hours after an eruption by Vesuvius. In 1795, another visitation, in the form of a fall of stones ten yards away from a startled labourer at Wold Cottage, near Scarborough, in England, provided a second endorsement of Chladni's conclusions, and a fall of stones out of the sky at Krakhut, in India, came in 1798. This fall was of the utmost importance in stifling scepticism, for the associated bolide or fireball arrived out of a clear, blue sky. Thus theories of condensation from clouds of ash and dust mixed with pyrites (still popular), and the perennial thunderbolt theory, became untenable. The visitation out of the clear, blue sky of Krakhut sounded their death knell.

Analyses by Edward Howard and mineralogical studies by the Comte de Bournon showed that the stones recovered at Lucé, Ensisheim, Wold Cottage, Krakhut, and Siena, and also one from

22

Tabor (Bohemia), were all of a single peculiar variety, quite unlike any known terrestrial lithic material because of the content of malleable metal. Yet scepticism still lingered on, particularly in France: where strangely it was soon to be rudely silenced by the fall at L'Aigle of no less than three thousand stones out of the sky. These were subjected to exhaustive study by Biot, who recognised the essential characteristics of *bolide* or *fireball* phenomena, so familiar to us today. He also noted the production of a *ground ellipse of dispersion*, with axes measuring 10 and 4 km—the clear imprint of a projectile spray impinging at an oblique angle.

From this time onwards meteoritics advanced steadily, both from an astronomical standpoint, embracing the behaviour of the mass coming in through the Earth's atmosphere and on its prior orbital journeying in space, and from a geological (mainly petrological) standpoint, involving description of masses recovered after observed falls or found by chance. Unfortunately, the study of such masses was long regarded as something of a museum curator's oddity, a sideshow of geology of little relevance to the mainstream of geological thought. Only in the middle of this, the twentieth, century was great attention focused on terrestrial and cosmic geochemistry, on terrestrial and cosmic physics, and the 'Space Age' concept took on a popular appeal, lunar and planetary exploration becoming feasible as a development of military rocketry. Meteoritics came to the forefront spectacularly, with countless scientists pursuing seemingly endless possible ramifications that rapidly became apparent, with the aid of ingenious and sophisticated instruments and methods, but, alas, perhaps not always backing their researches by truly scientific arguments. At present there is a lull in the space programme, after the Apollo and Mariner 'spectaculars', but there is no reason to doubt that space science, among other things embracing all aspects of meteoritics, is here to stay.

Meteoritics Today

To geochemists and physicists meteoritics affords the possibility of a backwards glance, discerning, if ever so dimly, the imprint of processes that occurred well beyond the three thousand nine hundred million year limit of our possible retrospection in observing the rocks that provide us with the terrestrial geological column (which is made up of a sequence of eras and epochs). The greater ages were estab-

23

lished by measurements of the decay of radioactive isotopes, by methods also applied to terrestrial rock material—though some entirely new methods have been developed purely for meteoritic studies. The first great leap forwards, outside the bounds of this planet, was obtained by the geochemist and physicist through the study of meteoritic material: the second such leap forward came with recovery of actual material for study from the surface of the Moon, and the third leap, involving recovery of material from the surface of Mars, may not be so far off.

The geophysicist supplements inferences directly derived from studies of seismic waves passing through the Earth and other terrestrial phenomena by studying meteoritic material, and, likewise, the geochemist supplements inferences derived from experimental studies and isotopic abundance measurements on terrestrial material by similar studies. Both working in complement have deduced models for the internal shell configuration of our own planet, based partly on studies of meteoritic material. Such material was the only extraterrestrial material available for direct study up to the Apollo and Luna recoveries from the Moon, and is supposed to represent the interior of asteroidal parent bodies, which may possess some degree of internal similarity to the Earth. Opinions differ, however, among scientists as to the number and size of such parent bodies—asteroidal parents that are believed to have been subsequently disintegrated by some obscure agency and flung in fragmental form as meteoroids into varied and commonly highly eccentric orbits. The evidence proliferates, and so do the conflicts of evidence, and nowadays, the practice of setting up terrestrial models based on meteoritics, a long established practice in geology, is becoming suspect. The aggressive, cock-sure scientist may appear entirely confident of his models and conclusions, yet the truth would seem to be that each new advance in meteoritics and planetary science yields new conflicts of evidence, and we are as yet little closer to achieving clear, unequivocal answers to the problems of the solar system. It also seems true that until we do resolve the problems of the originating processes for the solar system, by the study of meteorites and rocks from other planets, much will remain obscure about the deeper processes and remote beginnings of the objects studied by the relatively well-defined science of geology. For eruptive and metamorphic processes, and also geotectonic processes, hinge fundamentally on these deeper seated and initiating processes.

24

Meanwhile, in our state of clouded vision, we must retain as a working hypothesis the concept that in the iron meteorites we are seeing material analogous to the mysterious, part-molten core of our own Earth, the source of the Earth's magnetic field with its bewildering secular variations: also that we are seeing in the stony iron meteorites material analogous to the boundary zone between the Earth's core and major silicate layer, called the Mantle: and that in stony meteorites we see material analogous to that of that Mantle— even though we draw these analogies in full knowledge that the meteorite parent body may have been so much smaller in overall dimensions that its internal structure was widely different from that of our Earth. Nowadays it is becoming more and more apparent that meteorite material was formed under confining pressure regimes low compared with the internal regimes of the deduced internal shells of the Earth, and these analogies do appear to be somewhat dangerous. At present, however, we can draw no better analogy and it is doubtful if we shall ever see and be able to study directly the interior stuff that makes the large, terrestrial planets, for only the small meteorite parent bodies appear to have suffered disintegration.

Every time we handle a meteorite, or just gaze at it, we are likely to feel the shiver of excitement, of the mystery of reaching out and making contact with the unknown. And therein lies the importance of meteoritics: despite predictions that direct planetary recovery of material will supersede meteoritic material, meteorites are the only solid material that comes to us out of space, from the unexplored wastes enveloping our planet. This material comes to us continuously and at no expense: unlike recovery from lunar and planetary surfaces, immensely costly and gleaning only the surface cover, this material comes at no cost and represents the entire parent body, or so we believe. Further, it comes to us continuously; the store of meteorites is never complete, the book of meteoritics is never fully written: any day a new type of meteorite may come to us, representing an entirely unguessed facet of the solar system. Nothing can replace this wonderful and fortuitous manna from heaven to the scientist. The planetary recoveries certainly must not be discounted—they are of immense significance—yet the nature of the Moon and the nearest small planet, Mars, manifestly devoid of fold ranges to thrust up once-deep rock formations or erosive agencies to lay them bare, must have a monotony and a very incomplete representivity compared with the meteorite store.

25

The Scope of This Book

In the succeeding pages a picture is given of the store of knowledge about meteoritics that has come to us since Chladni and Biot dispelled the clouds of scepticism and demonstrated that lumps of metal and stone do fall from the sky. Starting with a consideration of comets, asteroids, and the meteor families in their orbits, we shall follow the latter down to Earth and then take up the story of the various fireball phenomena accompanying their fall, as seen by Earth-bound Man. An account of the physical nature of the masses left on the ground or in the ground, their grouping and their imprint patterns follows. The giant meteorite explosion craters are discussed later, in special chapters, alongside the obscure, catastrophic, unique Tungus event, and the enigma of the cryptic craters and crypto-explosion structures, thought by some to be 'astroblemes' or impact-scars, and by others to be of cryptovolcanic origin. Petrographic definitions and classifications are given for all meteorite varieties, and the minerals known to occur in meteorites are listed. Further special chapters deal with the implications of studies of radioactive isotope decay ('meteorite ages'), geochemical abundances, and the origin of chondrules—leading on to discussion of the origin of meteorites. The carbonaceous chondrites are covered, also, in a special chapter in the context of possible life forms. The possible relation between the crater-pocked lunar and Martian surfaces and meteorites provided the subject-matter for another special chapter. Tektites—glassy objects, believed by many to be a form of meteorite and certainly immediately projected from out of the sky, though their ultimate provenance provides science with one of its outstanding enigmas—are finally dealt with. These objects do not arrive continuously like meteorites, but sprayed the Earth during a few brief and geologically very recent periods.

A chapter, in the form of an epilogue, summarises the author's conclusions and predictions in the light of the Apollo and Mariner evidence.

3 : Astronomical

The Minor Members of the Solar System

The comets, minor planets (asteroids), and meteoroids, and also the fine dust responsible for the phenomenon known as zodiacal light, comprise only a small fraction of the total mass of the solar system. The term 'meteoroid' embraces not only the large masses which produce fireball phenomena in the day or night sky, and occasionally drop, fragmented or as single masses and considerably ablated, to Earth, but also the minute masses that drive at high velocities into our atmosphere producing the familiar 'shooting stars' of the night sky, leaving no material product in the form of meteorites on the surface of the Earth. It cannot be denied that these three types of minor members are related to one another, but scientists have only lately had some glimmering of the manner in which they come to be related. It must be emphasised that evidence of orbital coincidence in no way establishes genetic relationship, ie, that they share an ultimate provenance. Orbiting masses of quite different ultimate source could assume coincident orbital characteristics purely on account of a shared over-riding domination by some planetary perturbation.

Comets

The astronomer, J. G. Porter, notes that whereas an earlier worker, Crommelin, estimated the average period of comets at some 1,200 years, from which value it might reasonably be accepted that more than 120,000 cometary bodies exist, mostly following narrow, elongate and elliptical orbits extending out to some 1,200 astronomical units (1 au = mean distance from Earth to Sun, 93×10^6 miles or 150×10^6 kilometres) from the centre of the solar system, his figures were in fact based on frequency of *visual recognition* of comets possessed of such narrow, near-parabolic orbits, averaging three per year. There must be innumerable comets whose perihelia are situated so far out from the sun that they never become visible and never assume cometary characteristics familiar to the layman. Thus, there must be far more comets than Crommelin supposed, and Porter suggests extrapolation of populations within certain ranges of perihelia distances as a basis for computation of the likely number of comets: such extrapolation leads him to a figure of 10^{11} comets following orbits which take 40,000 years to complete on an average. J. H. Oort goes even further, envisaging a cloud of comets extending even further out into interstellar space, up to 200,000 au, and liable to orbital perturbations by the closer stars. The mind boggles at the population required by this theory, yet there are published records of only some 800 comets!

Such figures remind us that comets certainly journey far beyond the most distant known planet, Pluto (40 au from the centre of the solar system), and yet, in spite of these immense journeys out into space away from the Sun, comets do appear to be members of the solar system, not random visitors from the depths of space as was once believed. Porter concludes that comets may pass right out of the solar system on hyperbolic orbits, because of planetary perturbation, but that they do not ever initially enter the solar system on such orbits.

In addition to *long period comets* journeying on such immensely long, elliptical orbits, there are *short period comets* with periods ranging from three to one hundred and sixty years. Their orbits are still elliptical, but show less eccentricity and so more closely resemble those of the planets. They possess such orbits because they have suffered perturbation by the planet Jupiter, resulting in capture within the near-in part of the solar system while continuing to orbit the Sun, not Jupiter, which, the largest of all the planets, exerts by far the greatest perturbing influence of any body within the solar system. In the extreme case we see comets like P/Swassman-Wachmann 1 (comets are conventionally named after their discoverer, asteroids haphazardly, and meteorites after their place of fall or find), possessed of an orbit of extremely low eccentricity and thus remaining at an almost constant distance from the Sun, in orbit between Jupiter and Saturn, and also never at any time during their orbital cycle developing a significant tail feature because they never approach sufficiently close to the Sun. Such comets will be difficult to distinguish from minor planets or asteroids, just as some asteroids possessed of highly eccentric orbits (eg, Icarus) are difficult to differentiate from comets.

Yet the resemblance is superficial and astronomers have no real uncertainty as to which is which, for comets and asteroids have quite different celestial movements dependent on their mass and volume. Comets are bodies of very low total mass and according to the ideas of Whipple, now very widely accepted, the nucleus is small and consists of a 'conglomerate' of ices and dust. The 'coma' or bright halo develops only when the comet comes close to perihelion and is affected by the degree of closeness to the Sun: streamers of bright material, ionised gases and dust, trail out away from the Sun behind the comet as it passes through perihelion, forming the familiar tail. Comets can, however, appear to possess sunwards-directed tails— 'anomalous tails'—which are due to the observer's angle of view of the dust trailing out behind the comet in its orbital path. Such trails of dust, exemplified by the 'spike' of comet Arend–Roland, 1957, may have a close connection with the phenomenon of shower meteors. There are, however, other types of anomalous comet tail multiplications that appear to be real multiplications, not tricks of vision as in the case of Arend–Roland, 1957.

Spectrographic evidence suggests that elements of low atomic number—hydrogen, carbon, nitrogen, oxygen, etc—make up the greater part of comets. There are reports of detection of iron and nickel as well, and it seems possible that both metallic and silicate components may be disseminated within the ices. One must conclude that the material of the comets is not likely to make up the solid metallic and lithic masses which comprise meteorites without suffering radical differentiation involving loss of the volatile light elements that predominate, and are now held as ices, and concentration of the dust content into substantial secondary bodies. Even the supposedly primitive *carbonaceous chondrites*, meteorites which do show some chemical compositional similarity to comets, contain olivine, a high-temperature mineral not reasonably to be expected in cometary material, which seems to be essentially a low-temperature assemblage. A phase of high-temperature metamorphism would be needed to bridge this gap.

Asteroids

Asteroids possess orbits of less eccentricity than most comets, and they also have less orbital inclination to the plane of the ecliptic, the general plane of the solar system. It is true that twenty-two asteroids do possess orbital inclinations averaging 16° to the plane of the ecliptic, but these are still distinguishable from comets in that their motions reveal them as bodies of much higher density. The named asteroids number about 1,700 and range from a handful of large minor planets discovered with old-fashioned telescopes—Ceres, the precursor, diameter 770 km, and her lesser sisters Pallas, Vesta, and Juno, together with some later additions—to objects but a few hundred metres in diameter. The majority are of very small dimensions. Their motions suggest lumps of metallic and lithic material, and there is some indication of irregular rather than globular shape, in evidence of phases: this evidence supports the idea of fragmentation of a large planetary parent, an idea which may be called the classic theory of asteroids. According to the Titius–Bode astronomical law there is a missing space between Mars and Jupiter which should be occupied by a planet, and the possibility of fragmentation thus appears not unlikely. Yet the Russian astronomer, Schmidt, has proposed a much more attractive mathematical law for the spacing of the planets, a law which involves accepting that there are two groups of planets and that these exist because of a critical temperature boundary situated at about the radius of the asteroid belt, in the primitive circum-solar gas/dust nebula. The theory of Schmidt supposes the asteroids to be lithic bodies, remnants of the smaller planetesimal primary masses from which the planets were accreted in the cold state (remnants of such size that they could survive at this radial distance from the centre of the nebula, whereas a larger planet could not): the comets are supposed to be another form of such primary object, but formed and capable of survival further out in the nebula, where they are able to include volatile ices in their make-up, unlike the asteroids.

The comets and asteroids must have suffered severe planetary perturbations to possess their present orbits. The latter do, for the most part, orbit between Mars and Jupiter, but some come right in to perihelion close to the Sun—Icarus, a very small fellow no larger than a few city blocks, scorches his wings only eighteen and a half million miles from the Sun, well inside the orbit of Mercury: he thus also passes through the orbits of Mars, Earth, and Venus on this perilous passage, allowing for a real possibility of collision with a planet— albeit, he was sixteen times as far away as the Moon on his last passage in 1968 and so the dangers of Icarus tend to be over-sensationalised. Other asteroids such as Hidalgo orbit far enough out to intersect the orbit of Saturn. Astronomers have declared that there is a statistical likelihood of frequent past collisions between asteroids and the

Earth throughout the 3,500 million years covered by the geological column, but scepticism concerning the validity of this conclusion may be justified: only a handful of asteroids out of the 2,000 known possess orbits allowing any chance at all of such a collision, and there is a possibility also that the number of scars on the Earth's surface supposedly produced by giant meteorites is taken into account in this computation, and, as we shall see below (Chapter 20), such attributions are probably vastly exaggerated. The evidence of the extreme age of the rocks of Mare Tranquilitatis obtained by Apollo 11—a Mare of supposedly very youthful lunar comparative age—has pushed back the time of the supposed lunar bombardment to a date prior to the entire geological record: it seems probable that this bombardment is a figment of Man's imagination and that the cratering record is in fact an artifact of a primitive intense eruptivity, not extralunar bombardment.

Shower Meteors

The orbits of periodic comets—those of short orbital length imposed by capture by the planet Jupiter—and those of the asteroids are unquestionably related to one another. And the orbits of asteroids show a close relationship to those of groups of meteors which provide spectacular pyrotechnic displays in the night sky, 'shooting' stars radiating from specific points in the night sky (radiants) and recurring at distinct periods in the year or at distinct dates after intervals of a specific number of years. These 'meteor showers' are known by the name of the constellation in which the radiant appears to be situated, for example Leonids (Leo), Draconids (Draco), and Perseids (Perseus).

In 1865 the orbit of Comet 1862/III was recognised by the astronomer Schiaparelli to be closely related to that of the Perseid shower meteors. With Leverrier and Peters, the same astronomer also recognised shared orbits between Comet 1861/I and the Leonids. Weiss later extended this relationship further, noting that the Lyrids appear to share an orbit with Comet 1861/I, and, even later, it was demonstrated that the November Andromedids share an orbit with Biela's comet. Such observations were at first taken to indicate that comets and meteors are manifestations of a single phenomenon, of which meteorites are the only material survival reaching the surface of the Earth, and that both are derived from interstellar space, outside the solar system. This belief was in accord with the idea that meteor

31

orbits are hyperbolic, bringing them in from interstellar space and out again never to return, unless captured within the solar system by planetary perturbations. More accurate determinations have since revealed that it is unlikely that comets *ever* enter the solar system on hyperbolic orbits: and that meteors, both of the high-velocity/low-mass shower-meteor variety and of the low-velocity variety, which may have considerable mass, produce fireballs, and drop meteorites, must possess elliptical orbits compatible with an origin as members of the solar system.

Such orbits have been derived for the Pribram, Czechoslovakia, meteorite of 7 April 1959, by Cepelecha, and for the Sikhot-Alin meteorite of 12 February 1949, by Fesenkov. In both cases the orbits are of low inclination to the plane of the ecliptic, and resemble those of the majority of asteroids rather than those of comets. Cepelecha's computation was the first based on photographic evidence: some authorities have questioned the conclusions, believing that when possible perturbations are taken into account a lunar rather than an asteroidal source is indicated. Despite these objections, further computations of orbits from visual records (as in the case of Sikhot-Alin) are supporting the idea of asteroidal provenance—for example, the fireball of 7 June 1969.

The erroneous belief that meteors follow hyperbolic orbits may have stemmed from a total reliance on visual observations, which tend to be very inaccurate: photographic methods of recording now in use eliminate this inaccuracy, but, to date, relatively few fireballs that are known to have dropped a meteorite have been recorded in this way. In contrast, high-velocity meteors have been extensively studied photographically, and reveal no orbits that are not elliptical. The idea of hyperbolic meteor orbits is still favoured by Lapaz, though disputed by Krinov, and certainly not nowadays held in general favour by astronomers.

There is certainly as much astronomical evidence of coincidence of asteroidal orbits with those of shower meteors as there is for a comet/shower-meteor relationship. For example, the orbits of the asteroids Adonis, Apollo, and Hermes conform closely to those of the Virginid, Scorpio-Saggitarid, and Piscid meteor showers respectively. The truth would seem to be that we have a trinity, periodic comets, asteroids, and shower meteors, sharing orbits: a fact that does not in itself establish them all as originating in the solar system, for there remains the real possibility of secondary capture.

32

However, other lines of evidence suggests that all these astronomical objects are original members of the solar system.

There would seem, at first sight, to be difficulties in relating meteors to both comets and asteroids, within the primary framework of the solar system, for comets and asteroids are quite diverse entities. These difficulties are probably, however, not insurmountable. Comets have a very light mass in their small nuclei, while asteroids appear to consist of relatively dense masses of varying dimensions, minor planets up to several hundred miles in diameter, and probably lithic/metallic in composition. Unfortunately, very little, if anything at all, appears to be known about the density values for individual asteroids (could not some astronomer derive an accurate method of working out the density of an asteroid, such as Icarus, which passes very close to the Earth?). The common types of meteorites, lithic, metallic, or lithic/metallic masses of high density, which drop from fireball meteors, could not conceivably come from comets, which are known, from spectrographic evidence, to be composed largely of low-atomic-number elements, and are likely to consist of dust bound together by ices of the volatile components.

On the other hand, meteorites could easily be derived from the apparently dense material of asteroids. The lighter, high-velocity shower meteors could easily stem from cometary material, strung out behind the nucleus along the orbital path as the comet approaches the Sun, and displays its spectacular tail. Jacchia has concluded that these meteors reflect fragile masses, and upholds the ice/dust cometary model of Whipple, basing his arguments on studies of the deceleration rates as they drive into the Earth's atmosphere. He is able to recognise orbital stragglers from any particular group of shower meteors by virtue of this property. He has suggested that short period comets, held in secondary capture by planets, and following modified orbits due to the over-riding control of the planets, disintegrate faster than long period comets; and thus, in the case of the shower meteors related to these, the material comes from a resistant core representing the cometary nucleus. Other shower meteors are related to long period comets, orbiting immense distances out into Space beyond the limits of the planetary system of the solar system known to us, and thus not continually depleted by continuous, short period orbiting of the Sun in modified orbits: and so these consist of friable material, representing an outer layer sheathing the cometary nucleus, a layer no longer preserved in the case of the short-period comets. Thus, very fragile meteors, entering the Earth's atmosphere with high velocities, are related to long period comets, whereas the less friable, low-velocity entries are related to captured comets with short period orbits of regular character, resembling those of the asteroids–meteor groups most easily distinguished by the return of the shower within the span of a man's lifetime, at regular intervals.

The conclusions drawn by Whipple and Jacchia are attractive: one difficulty involved in acceptance of this theory is, however, the fact that it invokes two quite distinct types of meteor, and one would logically expect astronomers to be able to draw a sharp line of distinction between the two types of meteor showers. Attempts to do just this appear to have met with scant success. It seems logical that with increased brilliance of the meteor, a large percentage of the denser type of meteor, related to the short period comets and asteroids, should be expected. Suggestions have been made that all photographically detectable, radar detectable, and visually detectable meteors, other than fireball meteors brighter than the full Moon, stem from cometary dust. Whereas fireballs of lunar brilliance and some others which arrive with very slow cosmic velocity, and so dig down deep into the Earth's atmosphere before becoming luminous, are supposed to stem from asteroids. But, when attempts are made to go beyond this very arbitrary artificial division, it becomes evident that the case is not so clear-cut as this division implies. Data from orbits of 413 very bright meteors studied by Jacchia indicate that no more than 1 per cent could conceivably stem from asteroids, and the conclusion is inescapable that most very bright meteors are not of asteroidal origin. This is a field of research that could lead to much more positive results with the development of even more sophisticated methods of study.

Spectrographic evidence from shower meteors suggest that in chemical composition they are not unlike the material of stony meteorites, but this does not provide proof of either a cometary or an asteroidal source. The spectrographic evidence from comets suggests not dissimilar components in the nuclei. Especially significant is evidence of nickel and iron, which are, of course, major components of virtually all types of meteorites.

Summing up the astronomical evidence, it seems highly probable that the meteorites (masses that actually arrive on the surface of the Earth) are related to asteroids, being smaller fragments from the same parent body. A derivation from the dense, small nucleii of comets, of some meteorites (? carbonaceous chondrites types 1 and 2) cannot be ruled out on astronomical grounds, though the petrographic study of the meteorites themselves suggests that they are essentially congenetic and, therefore, that this is extremely unlikely. The astronomical evidence tends to favour a monistic theory, in which comets, asteroids, and meteors are all ultimately derived from a reasonably uniform parental nebula source material, and all are members of the solar system: the model for the derivation of the planets and minor members of the solar system by accretion in the solar nebula (of Schmidt and others) seems, broadly speaking, to fit in well with the accruing astronomical evidence, though it needs some modification with respect to the mode of derivation of the planets and asteroids, and, suffers from a weakness in that the small proportion of the angular momentum of the solar system taken up by the Sun has yet to be explained adequately: despite this it provides by far the best model for use as a basic assumption in discussions of the origins of meteorites.

The biggest problem facing the astronomer would seem to be explaining the derivation of the meteorite-dropping meteors from the immediate parent body: this is, of course, the problem of derivation of the asteroids with their strange orbits from their original parent body, or bodies, orbiting between Mars and Jupiter. Fragmentation of a minor planet, explosively, is not easy to explain, though of recent years it has been suggested that planets could develop an increasing metallic core to the point of instability, at which point they might 'blow themselves to pieces'. But such ideas are only in their infancy. Whatever the reason for the fragmentation, we shall see, when we come to discuss the origin of the meteorites, there seems to be little doubt that they represent fragmented parent body detritus. The size of the meteorite parent body or, as is now becoming almost certain, several parent bodies, remains a matter of controversy—asteroidal, lunar, or planetary size?—all have their advocates. It has been supposed that the fact that the known asteroidal material cannot account for more than 0·1–0·2 per cent of the volume of the Earth severely restricts the possible dimensions of the parent body, or bodies, but this argument may well be fallacious, for much of the product of fragmentation could have been lost by spiralling into the Sun. The best reason to discount a planetary sized parent body seems to lie in the evidence of cooling rates for meteorite metal alloys (Chapter 19): these certainly seem to indicate asteroidal dimensions.

The number of parent bodies prior to fragmentation is again debatable—some authorities invoke a number of bodies in the parent body phase, later disrupted and thrown into eccentric orbits, a few to tangle with the Earth: thus Mason has lately come to favour a model including some of large asteroid (= Ceres) size, and a gradation from these down to 'football' size. There are, however, strong objections to the proliferation of parent bodies, for there is much evidence that many of our known meteorite types stem from the same individual parent bodies (evidence derived from polymict breccia associations and heterogeneity in the irons). As yet, no satisfactory model as regards numbers of parent bodies seems to have been derived, but the limitation of large asteroid size as the upper limit for the parent bodies seems well established.

4: Meteorite Flight, Falls, and Impacts

Behaviour of Meteoroids in Flight through the Earth's Atmosphere

Related to the velocity of meteors, and their mass and fragility, is the variation in the altitude above the Earth's surface at which the onset and termination of luminosity occurs. The trajectories of meteoroids have been discussed particularly fully by Fedynsky and his conclusions are summarised below.

The height of appearance or onset of luminosity (H_1) and the height of the disappearance or termination of luminosity (H_2) are the principal terms used, but the height of the middle of the luminous path (H_0) is also a useful term. Methods of study are visual observation (naked eye, telescope), photographic, and radar. The latter method depends on the condensation trail, not on direct measurement of the luminosity and an absolute agreement of values obtained by this method with those obtained by the other methods is not to be expected, but agreement is sufficiently good for this method to be a valuable adjunct. Radar also suffers from the limitation that it cannot distinguish between H_1 and H_2 for any but very bright meteors (greater than magnitude 6).

Values for 'photographic' and 'bright' meteors (-5 to 2 magnitudes) are:

$$H_1 \quad . \quad . \quad . \quad 110\text{–}68 \text{ km}$$
$$H_2 \quad . \quad . \quad . \quad 100\text{–}55 \text{ km}$$

For 'bright' meteors (greater than magnitude 6) detected by radar they are:

$$H_1 \quad . \quad . \quad . \quad 115\text{–}100 \text{ km}$$
$$H_2 \quad . \quad . \quad . \quad 85\text{–}75 \text{ km}$$

And, for 'faint' meteors (less than magnitude 6) detected by radar:

H_0	.	.	120–111	110–101	100–91	91–81	80–71 km
Percentage	.	2	25	37	24	12	

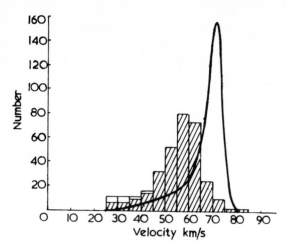

Fig 2 *Velocity distribution of meteors according to radar observations. The smooth curve gives the distribution expected in the case of parabolic meteor velocities. The rectangles are observed elliptical velocities (after V. V. Fedynsky)*

Fig 3 *Heights (H) of meteoric phenomena. Left, limit heights: the beginning and end of paths of bolides (B), visual meteors (M), and radar meteors (RM), telescopic meteors in visual observations (T): M_T is the delay region of meteorites (after V. V. Fedynsky)*

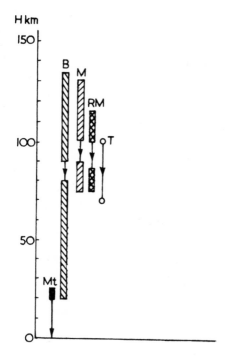

Observation heights mainly extend from 120 to 70 km, and faint, low-velocity meteors, visible with telescopes as they glow after digging deep into the atmosphere, appear to be luminous at even lower altitudes. They do not glow until past the 70 km level and reach their maximum luminosity between 60 and 40 km according to the Russian astronomer, Astapovich.

The very bright fireball meteors, bodies of considerable mass, glow at even deeper levels but are luminous throughout a long trajectory, as is illustrated by their values:

$$H_1 \qquad . \qquad . \qquad . \quad 135\text{--}90 \text{ km}$$
$$H_2 \qquad . \qquad . \qquad . \quad 80\text{--}20 \text{ km}$$

Fireball meteors which achieve penetration to beyond 55 km without complete dissipation produce sound effects, while those that penetrate to 25–20 km actually drop meteorites, and *are the only meteors*

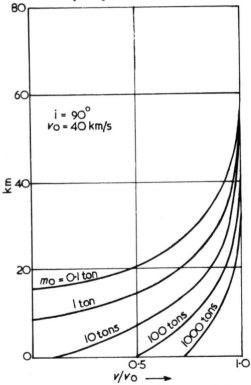

Fig 4 Diminution of velocity of meteorites of different mass during passage through the Earth's atmosphere, assuming an initial velocity of 40 km/s and vertical infall (after F. Heide)

that do. It is important to note that not all fireball meteors drop meteorites, hence many prolonged and abortive searches after fireball observations. In many cases a highly spectacular display in the sky is followed by no actual access of material to the Earth's surface. Altitude ranges of various meteoritic phenomena are well summarised in two figures given here, after V. V. Fedynsky.

The fireball meteor, if it does drop a meteorite, does not of necessity drop it with any significant velocity. It is commonplace to hear talk of meteorites falling with immense velocity and indeed it is difficult for the non-scientific among us to imagine anything else, taking into account the immense orbital velocities. Such a high velocity of impact is, however, only inevitable on an atmosphereless surface, such as that of the Moon, for the Earth's atmosphere brakes the meteorite in its fierce, supersonic approach, and most meteorites pass through a point at which all cosmic velocity is lost, after which they come towards the Earth's surface at merely free-fall velocity, pulled down solely by the Earth's gravitational field. Thus, it is possible for even the largest known meteorite, Hoba, South West Africa (sixty tons), to be found only partly buried in the ground. The velocities and trajectories of meteorites are shown in a figure after Heide, assuming an initial velocity of 40 km per second on reaching the fringe of the Earth's atmosphere, a velocity that is unlikely to be exceeded.

The existence of certain well-authenticated large meteorite craters does suggest that there are rare oversized visitors—large masses which do come in with high geocentric velocities; that is, retaining part of their cosmic (orbital) velocity. Cosmic velocities may vary, and an unusually high cosmic velocity would increase the chances of high velocity impact. Yet we believe that meteors are members of the solar system and there is thus a limit of 42 km/s to the velocity of meteoroids in space—above this value they would escape from the solar system. Whereas retrograde orbits and head-on collision with the Earth, adding the Earth's orbital velocity (30 km/s) to that of the meteorite, are theoretically possible—taking the maximum initial velocity at the time of entry into the atmosphere up to a possible value of 72 km/s—there are strong arguments against these retrograde orbits being followed, and it is doubtful if any meteorites have initial velocities exceeding 40 km/s, while most have much lower initial velocities.

It may be argued that some shower meteor families have retrograde orbits (eg, the Leonids), and therefore meteorites may well prove to possess such orbits. There is, however, apparently no indication of this from computed meteorite orbits; to invoke such cases, as has at times been done, would seem to be tantamount to accepting other than asteroidal (ie, periodic comet) sources for the meteoritic detritus, that is non-asteroidal parent bodies.

The actual mass and the angle of approach are the factors that control high speed impact—only a very large mass can bore through the atmosphere virtually unbraked, by virtue of its momentum, overriding atmospheric opposition: and approach in a vertical path—from the zenith—will mean that less atmosphere will be encountered before impact, and the braking effect will thus be less.

The fact that the meteoroid is not approaching a stationary planet must not be overlooked. It has been popular to suggest head-on collision with the Earth as a real possibility (giving addition of cosmic and terrestrial orbital velocity—theoretically up to 72 km/s). Overtaking from the rear will, of course, involve terrestrial orbital velocity acting to reduce the effect of cosmic velocity and there is a theoretical minimum incidence velocity at the fringe of the atmosphere of 12 km/s. Radar determinations indicate geocentric velocities ranging between 15 and 70 km/s for bright meteors—the maximum provides a confirmation that meteors belong to the solar system and do not move in orbits of parabolic character at velocities in excess of 42 km/s right out of the solar system, never to return. In spite of these figures it seems that the head-on collision of meteorites is not likely; for we must reasonably connect meteorite-dropping meteors with the asteroids, and these, like all members of the solar system except comets and one or two anomalous planetary satellites, orbit in direct rotation. Further asteroid orbits (with only very few exceptions) are close to the plane of the ecliptic, only a small group of about twenty having orbits appreciably inclined to it, and their orbits do not exceed inclinations of 30–40 degrees. Now, even if such asteroidal masses do come through the Earth's orbit to approach closer to the Sun—and this is the only type of orbit that could allow collision with the Earth at some point on its trace—the asteroid could never collide with the Earth head-on, unless it possessed retrograde movement. The low inclination of asteroid orbits to the plane of the ecliptic and the coincidence of meteor shower orbits with those of asteroids suggests that meteoroids in showers related to periodic comets, and meteoroids related to asteroids, must have a reasonably

41

low initial velocity, probably never exceeding 20 km/s, at the time of atmospheric entry. The few higher initial velocities recorded by radar must be related to non-periodic comets.

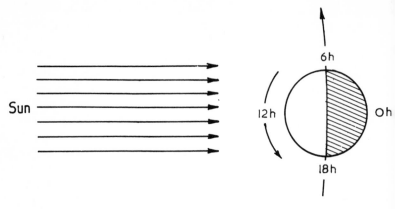

Fig 5 The variation in meteorite incidence geometry with time of day, ie, dependent on the rotation of the Earth on its axis (after B. Mason)

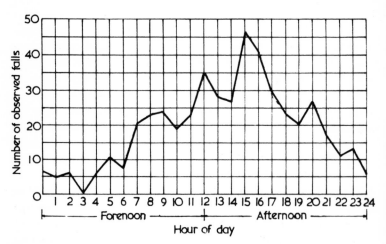

Fig 6 Hourly variation in the incidence of meteorite falls, 1790–1940 (after Leonard and Slanin, 1941)

The statistical variation diagrams of Mason are interesting, for they illustrate the control exerted on meteorite incidence by:

(a) The date in the year.

(b) The time of day.

The variation with time of year must be related to incidence of

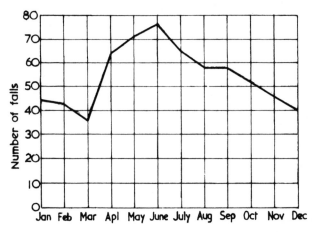

Fig 7 Monthly variation in the incidence of meteorite falls, 1800–1960 (after B. Mason)

meteorite swarms, orbiting in such a manner that, at a particular time of year their orbit tangles with that of the Earth. Recent studies suggest that certain meteorite incidences can be related in this way to shower meteorite occurrences, a fact that suggests that the shower meteors may well have a few larger masses travelling along in their midst, large lumps amongst a swarm of minute meteoroids. There is evidence that the shower meteors are very unevenly dispersed along their orbital paths and some Earth intersections may tap the main swarm, whereas others may be slightly out of phase and thus tap only a few stragglers or precursors.

Recent work also suggests that orbits worked out for the incidence of meteorites or fireballs (bolides) are also compatible with derivation from larger masses which are fellow travellers amid the shower meteor swarms, and ultimate derivation from the asteroid belt or from disruption of periodic comets is thus strongly indicated by both these lines of evidence.

Meteorite Falls Observed from the Ground

Fireballs, Dust Trails, Sound Effects

It has been recognised, in the previous chapter, that the little

43

'shooting star', the 'spark in the sky', does not drop a meteorite. It is significant that there is no increase in the number of falls recorded at the time of the periodic meteor showers, such as the Leonids: only a single record exists of a mass actually seen to fall during such a shower, in the case of the November Andromedids of 1885. During this shower 75,000 meteors were estimated to enter the Earth's atmosphere during a single hour. During the Leonid shower observed from Kitt Peak, Arizona on 17 November 1966, a rate of 60,000 per hour was maintained for 40 minutes. It seems obvious that such showers may never drop any solid mass of any appreciable size to Earth, and the one known case cited above may be a fortuitous coincidence: the mass may have had no genetic relation to the Andromedid meteor shower, though it may have shared an orbit with the shower meteors. This mass was an iron meteorite of normal type, and fell at Mazapil, Mexico.

Modern investigations of meteorite and fireball orbits do, however, suggest that some larger meteoroids, capable of reaching the surface of the Earth, travel along amid some shower meteor families, sharing the same orbits. Even then they may not be congenetic, that is sharing the same provenance with the shower meteors—as noted in Chapter 3, the association may be due to secondary perturbations bringing material of diverse provenance together in a shared orbit. Despite these reservations, one thing is certain—that it is the bright fireball meteor, brighter than the full Moon by night, lighting the night landscape momentarily like daylight and even appearing luminous in daylight in conflict with the Sun, that yields meteorites: but, even then, it seems that only a small percentage, possibly a fraction of 1 per cent, of these bright meteors do drop a meteorite mass. The term 'bolide' is commonly used for these visual accompaniments of meteorite arrival, and is, perhaps, better than the term fireball, for the latter term has also been applied to globe lighting, a quite unrelated phenomenon. 'Fireball' is also expressive of the nature of the phenomenon.

By night, a fireball meteor will appear slightly pear-shaped and may exceed the Moon in apparent diameter. The fireball flashes across the sky in an apparently straight trajectory, its visible passage lasting up to a few seconds. Its path may extend for hundreds of miles. At its maximum luminosity the country is lighted like daylight and this moment of maximum luminosity occurs when the mass is seen to explode; such explosions may occur once or twice during

Plate 1 The fall of the Sikhot-Alin meteorite in 1949. Painting by P. I. Medvedjev at the offices of the Meteorite Committee of the Academy of Sciences of the USSR, Moscow

flight and the ball of light either expands abruptly or separates into two or more fireballs. The colour in the early, hot stage may be bluish or greenish, but turns to pinkish with deceleration, and, in the very final stage, the incandescence wanes, the globes giving place to a trail of sparks glowing like embers. At night the cloud which is left as a trail behind the fireball, and which consists of ionised gases, appears luminous.

By daylight a fireball appears much less luminous, appearing as an indistinct silver streak, but it leaves behind a cloud of dust particles which interfere with the passage of the Sun's rays and so appear dark. Such clouds may persist for several hours, becoming progressively more distorted to a sinuous form by the winds of the upper atmosphere.

If the observer is at or near the eventual point of impact he will, at night, see a star-like object growing larger and larger, and passing

45

through similar changes, without giving the effect of streaking across the sky. He will be looking along the line of approach and thus sees no transverse movement. It is likely that an observer so placed may not have his attention drawn to the fireball until a comparatively late stage in its approach, and he may even report that there was no accompanying fireball.

The last stage in the approach to Earth is the fall of fragments (if the fireball actually does drop a meteorite). A single mass, or more often a number of fragments, separated from the original mass during atmospheric entry, now no longer even weakly incandescent, will be seen to fall under the influence of gravity at free-fall velocity, each fragment being followed by a trail of grey 'smoke'. The dust trail, evident as a dark cloud by day, commonly appears to terminate in a swelling, which produces the effect of a small, distinct terminal cloudlet, hovering right over the actual point of fall.

Sound effects are complicated, and James Thurber's version ('A huge pink comet, Sire, just barely missed the Earth a little while ago. It made an awful hissing sound like hot irons stuck in water'): an earlier description in the Lavrentevko Annals of AD 109, quoted by E. L. Krinov ('Vesevelod . . . saw a large serpent falling from the clouds . . . all this time the Earth was rattling . . .'): and Tennyson's imagery ('Now glides the silent meteor on . . .') all appear to have some scientific foundation. The shower meteor is soundless, and so too are many fireball meteors. The observer may hear nothing even in the case of very bright, meteorite-dropping fireball meteors, for there is a silent zone surrounding the area of impact where sound effects are intense: beyond the silent zone, at a distance from the area of impact, there is a further zone in which the sound effects may be picked up by the observer. The sound-effect zones are elliptical in shape in the usual case of oblique incidence (in the very special case of approach from the zenith they would be circular). Sound may be heard up to 80 km from the point of impact in the axis of flight, and up to 30 km on either side of the flight path.

If sound effects are not heard by the observer he is certainly not very close to the point of impact, and it is a question of fundamental importance to be put to an observer, 'Did you or did you not hear any noise?' The negative reply will probably mean that search in that immediate area is futile: the converse does not apply, unfortunately, for because of the existence of three elliptical sound-effect zones and the silent zone an observer who hears sound effects may be even further

46

from the point of fall than one who does not. Reports of falls which do include descriptions of sound effects are worth following up, but those in which no observer records any sound effect are seldom worth it.

The variety of the sound effects experienced is a feature of descriptions by observers of meteorite falls, but amid this variety a pattern of consistency does appear:

1 A slight whistling or cracking is recorded during the actual observation by many observers. This seems to be a quite genuine experience, though some authorities regard it as illusory. In the case of a meteorite fall in Africa, a native woman is reported to have removed her baby from her hut before the meteorite landed nearby, because she heard this warning sound. The origin of this sound-effect is obscure.

2 Loud reports, commonly single, or two or more in succession, follow a minute or two after the visual observation. These are analogous with the now familiar shock wave detonations produced by supersonic jet aircraft.

3 A noise like thunder, tearing of calico, or the passage of an express train over a viaduct usually follows after these detonations, which have themselves been likened to cannon fire.

4 Near the actual site of fall the sound of fragments splitting off from the original mass at the end of the luminous path may be heard. Whistling and buzzing noises like those produced by falling bombs are described, and, on occasion, the observer, if very close to the point of fall, may hear the actual thud of the mass landing.

A confusing fact is that the sound effects appear in reverse order to an observer near the point of impact—he hears the arrival sounds, then tearing noises, and lastly the detonations. The sound appears to move away from him along the approach path, the earliest produced sounds, the detonations, reaching him last. The observer directly beneath the luminous part of the trajectory will experience shock-wave detonations first: a careful interrogation can elicit from observers the chronological order in which they experienced the sound effects, and thus establish their position relative to the flight path and point of impact.

An observer of a meteorite fall or a very bright fireball should take a bearing with a compass or using a reference point from his own standpoint (which he should mark) on the point of disappearance of

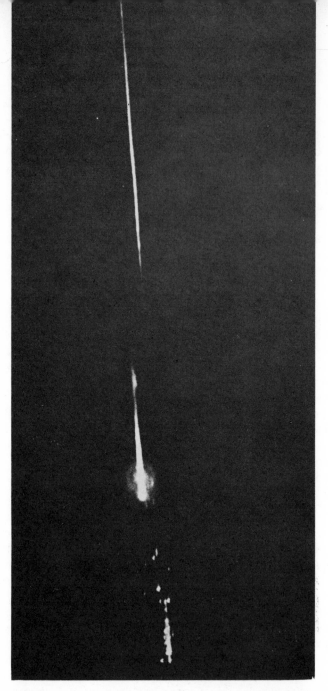

Plate 2 The re-entry of Apollo 13. The similarity to the meteorite bolide is striking, in this photograph of the re-entry of the separated command module and supporting unit at the end of the uncompleted mission plagued by near disaster

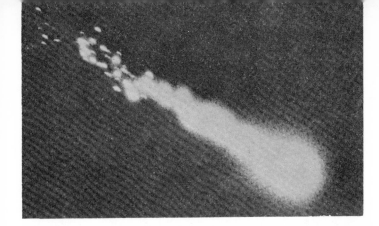

Plate 3 (top) Fireball of 25 April 1969, photographed by Mr H. M. Walker, from Bangor, North Wales, showing a multiple centred bright head and fragments separating at the tail. This fireball was responsible for the fall of the Sprucefield and Bovedy falls of stony meteorites in Northern Ireland; (bottom) Three progressive stages in the break-up of a fireball photographed on 25 April 1966 by Mrs V. Schwarz of New Jersey

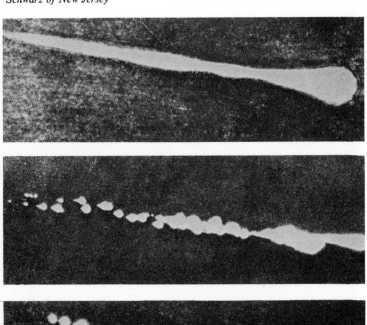

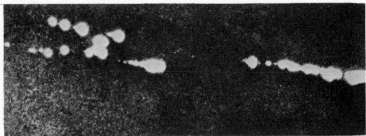

the fireball or apparent fall of the meteorite. He should also note the exact time of day (later checking his watch for accuracy). Finally he should try to reconstruct the observed path of the bolide across the sky, particularly with reference to the stars, if visible, making a sketch on the spot.

Impact Effects

Observations made of meteorite falls suggest that falls at very high velocities are extremely rare. Soft ground or soil cover may be penetrated to depths of about two metres, but, quite commonly, meteorites are found lying on the surface, not even partly buried. A meteorite fell at Hessle, Sweden, in January 1869, and the fragments came to rest on the ice surface without even cracking it! The recent find of a mass of about 12 tons near Forrest in Western Australia (the Mundrabilla meteorite), a find with which the author was associated, provides another excellent example of minimal penetration. The mass rests on the surface of a limestone desert, just as it landed, and shows no associated crater or penetration of more than eight or ten centimetres into the meagre top soil, coming to rest on the bed-rock, which shows some cracking but little other sign of disturbance. The Hoba meteorite, the largest single mass known, from south-west Africa (c. 60 tons), is barely buried in the soil. Small masses coming in at free-fall velocity may produce a small cavity in the soil, but this is little larger than the individual mass, and is an *impact pit* easily distinguished from a *fragmentation* crater, an open crater containing and surrounded by small fragments of the meteorite.

If meteorites do come in to the ground at high velocity they tend to fragment. Velocities of between 200 km/s and 4 km/s are supposed to be those limiting fragmentation effects, on impact. The effect of a hard surface—rock, hard pan, etc—and also the effect of inhomogeneities in the meteorite itself, such as a patchy texture or easy planes of fracture bounding large crystal growths, can assist the fragmentation. For example, the Mt Egerton meteorite, a unique meteorite found in Western Australia consisting of crystals of enstatite several centimetres long, aggregated like a coarse pegmatite (a coarsely crystalline feldspar rock familiar to geologists), has yielded nothing but fragments a few centimetres in diameter, because of the easy fracture supplied by the crystal boundary planes. Thus, a meteorite with an incidence velocity rather lower than the theoretical limiting value given above might well suffer fragmentation, because of hardness of ground surface or internal structural factors.

Impact fragmentation craters are a feature of arrivals involving intense fragmentation. These are typified by the renowned craters produced by the great iron meteorite fall at Sikhot-Alin, Siberia, in 1949. Krinov describes 122 craters arranged in the form of an ellipse, and varying from 10 metres diameter to 26 metres. Other craters of

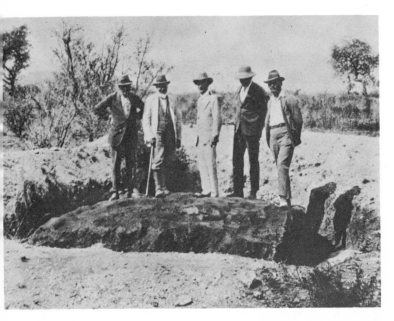

Plate 4 The World's largest single meteorite mass. The Hoba meteorite, South West Africa, showing the shallow penetration into the soil. The meteorite still lies at its original resting place

this type include the Kaalijarv crater, in Estonia—also described by Krinov, this is the largest of a group, and is 110 metres in diameter—and the Dalgaranga crater, Western Australia, accounts of which have been given by Nininger and Huss and the present author. This is a single isolated crater about 25 metres in diameter, with conspicuously up-arched country rock at the rim: it is 3 metres to the top of the fill and probably did not have an original depth in excess of 8 metres. Fragmentation craters tend to be surrounded by small, twisted masses of metal, and this is true of the Dalgaranga Crater,

51

although the meteorite has been shown to be a mesosiderite stony-iron with only scattered nodules of nickel-iron. The Boxhole Crater, Central Australia (described by Madigan), and the smaller of the Henbury Craters (first described by Spencer), also in Central Australia, also seem to be of this type, as do the Wabar, Arabia, Craters described by the same author: and in America there are the well-known examples of the Odessa Crater and the Haviland Crater, the latter associated with the Brenham pallasite (both described by Nininger): in Mauritania there is the El Aouelloul Crater described by Monod and, lastly, there is the Campo del Cielo Crater, Argentina (first described by Spencer). While there are probably a few others not listed here, this short list emphasises the rarity of such fragmentation craters of any size. All except the Haviland and Dalgaranga Craters are related to iron meteorites, which seem to orbit in larger masses, and also tend to fragment abruptly on impact because of the existence of major crystallographic partings in the mass. Only Sikhot-Alin represents an observed fall.

The impact fragmentation craters are well authenticated, and some much larger 'impact explosion craters', apparently due to vaporisation of very large masses coming in with some retention of cosmic velocity, have been proved, almost conclusively, to be of extra-terrestrial origin. Over and above these, however, there have been many very wild attributions to extra-terrestrial agencies, for both very large craters and crateroid structures miles in diameter, and other circular geologic structures, particularly cryptic structures associated with signs of explosive brecciation and little or no signs of magmatic emission. These attributions are covered in a separate chapter (Chapter 20), as a specialised topic related to meteoritics.

To conclude this chapter, we must make mention of *dispersion ellipses*. Meteorites coming in on vertical trajectories from the zenith, a rare case, will produce circular dispersion patterns if there are a number of fragments broken off in atmospheric flight. In the usual case of oblique approach the pattern of dispersion will be elliptical. Commonly, the fragments are scattered within the ellipse in a systematic manner, the largest masses carrying further to the forepart of the ellipse, and the medium-sized fragments falling in the centre, while the smallest fragments occupy the rear end of the ellipse. The larger fragments retain some cosmic velocity for slightly longer and push out beyond the smaller fragments in the direction of travel (see Heide's curve). This ideal pattern is not, however, by any means recognised in every ellipse, because of complications of repeated disruptions which impose a secondary scattering of fragments, and lead to ellipses with more or less random scattering of fragments. The effect of the wind can also complicate the distribution of small fragments, deviating them in the later stages of their fall. In the accompanying figure some recorded ellipses of dispersion are shown diagrammatically.

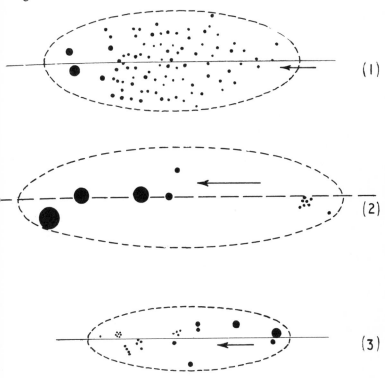

Fig 8 Dispersion ellipses. (1) Homestead, Iowa; (2) Kunashak, Siberia; (3) Johnstown, Colorado. The last was observed to approach as a bolide from a contrary direction to the usual one (note arrow). (1) and (2) are the normal cases (after H. H. Nininger)

The author has concluded from a recent study of the dispersion of the several hundred fragments of the Wiluna meteorite shower (2 September 1967), Western Australia, that direction of entry into an impact pit and bounce mark trend do not necessarily reflect orbital vectors and direction of approach. It seems that, in the terminal

53

phase of spiralling down of masses in free fall, the direction of travel of the bolide no longer controls the lateral movement of the mass. It is also interesting to note that there is evidence in this fall and another in Canada, that observers saw a bright object move across the sky in extension of the path of the bolide, continuing beyond the actual site of the find. Some form of 'plasma' may be responsible for this effect, becoming detached from the meteorites. However, part of the mass first fragmented high in the atmosphere may have continued on, to fall many miles away. It is now becoming apparent that the same arrival event can produce separate falls at quite widely scattered points. Up to now meteoriticists have regarded the extent of dispersion ellipses, seldom more than 16 km or so, as limiting the spread of contemporaneous falls. However, the Belfast events associated with a fireball seen over a wide area on 25 April 1969 provide critical evidence modifying this conclusion, for two separate masses fell at Lisburne and Kilrea 61 km apart, and the main mass or swarm appears to have continued on and fallen in the Atlantic, out beyond the Western Isles.

5: The Morphology of Meteorites

Meteorite masses are found on the ground in groups of fragments that may be either due to spread of fragments of a mass disrupted on impact or dispersion of a number of fragments which arrived simultaneously after disruption during atmospheric entry. In the latter case the group will consist of a number of individual masses each showing ablation effects, due to surface heating and melting during atmospheric entry. In the first case the masses will be irregular broken fragments, showing many faces without any ablation markings.

Meteoroids in all probability do not travel in very close groupings of masses in Space. What enters the Earth's atmosphere is a single mass. On occasion this may continue right in to impact without fragmentation. This will commonly produce what is known as an *oriented meteorite*, showing a distinct ablation pattern of anterior surface, from which material is melted off, and rear surface which collects this material in a thickened glassy skin. If, however, the shape of the mass is aerodynamically unstable, it may rotate during atmospheric entry and thus not show this distinct separation into anterior and posterior surfaces, producing oriented form. More often the mass breaks up explosively during its entry and gives rise to numerous smaller masses, each of which suffers ablation, developing its own fusion crust. If this fragmentation occurs late in the atmospheric entry there may be an original or primary surface of ablation, covered by fusion crust which shows numerous ablation markings, and the new fracture surfaces will be only moderately marked on the fusion crust, which may be very thin and only incipiently developed. Such surfaces have been referred to by Krinov as surfaces of the 'second kind'. Two phases of explosive fragmentation may occur during atmospheric entry in some instances. Also, some members of 'shower' arrivals may show pronounced orientation characteristics.

Plate 5 Anighito from Cape York, the largest meteorite mass in captivity—36·5 tons

Irons quite commonly continue in as large, solitary masses, and even when they break up, commonly break up into a few large secondary masses, often at a very late stage. However, some irons break up into an immense number of masses: the Sikhot-Alin meteorite is believed to have broken up in this way, only about 6 km above the surface. There seems to be quite a similar control in the case of the Mundrabilla Meteorite, Western Australia, found in 1966. Two masses weighing 12 and 5 tons are surrounded by an immense number of small iron fragments and these were found up to 24 km from the main masses, which were found 182 metres apart and represent a single mass which broke apart just before landing. The meteorite is characterised by very large and numerous nodules of the sulphide mineral troilite, which melts very readily, and appears to have continuously shed masses several centimetres long from its surface during ablation, as the surface became pitted and undermined due to the rapid removal of the support of the troilite component. Further investigation has shown it to be a brecciated octahedrite with silicate inclusions. Like the Sikhot-Alin masses, it consisted of an

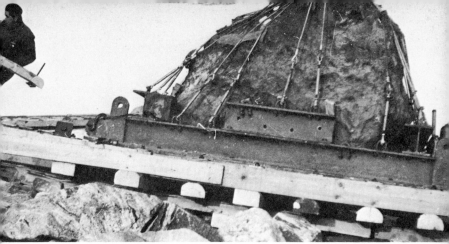

Plate 6 *Illustrating the problem of recovery of large irons: a 15 ton Cape York iron, Agpalik, being transported from the recovery site*

aggregate of 'fragments' with differing crystallographic orientations, so providing many planes for incipient disintegration to develop upon —in addition the troilite, carbon, and schreibersite banding along these surfaces in the Mundrabilla Meteorite will have been very readily removed.

Irons are found as much larger masses than stones or stony-irons. The World's largest known mass, Hoba, South West Africa, weighs 60 tons: it still rests where it was found, but Admiral Peary managed to bring back three very large irons from Cape York, Greenland, and these may be seen in the American Museum of Natural History, New York. They are called 'Ahnighito' ('Tent'), 'Woman', and 'Dog': the former rests on a weighing machine for all to see its weight, 36·5 tons! Another mass, 'Man' is in the Mineralogisk Museum at Copenhagen.

Plate 7 *A meteorite mass (iron) on display in Outer Mongolia. The mass is reported to weigh 35 tons and is reminiscent of the Willamette meteorite in its rearward surface, pitted in ablation flight through the Earth's atmosphere. It shows pronounced orientation features. It is presumably the Armanty iron*

TABLE 1

Large Irons

Hoba, South West Africa	60 tons	
Agpalik	20·1 tons	
Ahnighito	36·5 tons	The Cape York Greenland
Man	3·5 tons	Irons
Woman	3 tons	
Dog	400 kg	
Bacubirito, Mexico	27 tons	
Mbosi, Tanganyika	25–27 tons	(estimated)
Armanty, Outer Mongolia	20 tons	(estimated)
		(find of an iron reported by W. A. Cassidy in Argentina, September 1969, estimated at 18 tons)
Willamette, Oregon	14 tons	
Chupaderos, Mexico	14 and 7 tons	(estimated)
Campo del Cielo, Argentina	14 tons	
Mundrabilla, Western Australia	12 and 5 tons	
Morito, Mexico	11 tons	
Bendego, Brazil	5 tons	
Cranbourne, Australia	3·5 and 1·5 tons	
Adargas, Mexico	3 tons	
Santa Catharina, Brazil	2 and 1·5 tons	
Quairading, Western Australia	2 tons	
Chico Mountains, USA	2 tons	
Sikhot-Alin, USSR	1·7 tons	(largest mass)
Casas Grandes, Mexico	1·5 tons	
Navajo, USA	1·5 tons	
Magura, Czechoslovakia	1·5 tons	
Quinn Canyon, USA	1·4 tons	
Santa Apollonia, Mexico	1·3 tons	
Kouga Mtns, South Africa	1·2 tons	
Goose Lake, USA	1·2 tons	
Murnpeowie, South Australia	1·1 tons	
Zacatecas, Mexico	1 ton	

As against this long list, there are only two stony-irons known to weigh more than a ton—both pallasites.

Bitburg, Germany	1·5 tons
Huckitta, Central Australia	1·4 tons

The largest recovery of stony material is Norton County, an enstatite achondrite (1 ton): the recovery of the Allende, Mexico, stones after their fall early in 1969 is reported to have produced more

Plate 8 The Willamette meteorite, found near Oregon City, and weighing 15 tons. The basal surface is remarkably pitted, probably due to terrestrial weathering enlarging ablation pits such as those seen on page 57

than a ton of fragments of carbonaceous chondrite, type III material. A new find in 1968, north of Leonora, Western Australia—the Wildara meteorite, a common chondrite—totals an estimated weight of half a ton. Few other stony meteorite recoveries total much more than 500 kg weight, and such extensive recoveries are rare. Stony meteorites seem both to have formed smaller masses in Space and to have been more susceptible to fragmentation during atmospheric entry, and on earth impact. There are no large single stony masses of comparable size to the great iron masses.

Some very large meteorites may show perfect orientation: Mundrabilla certainly does, and Willamette, with its perfect cone form and deeply pitted base, is also regarded by many authorities as an oriented meteorite.

Very small meteorites down to a millimetre or so diameter are recorded from the Sikhot-Alin and Holbrook falls, and some of the numerous specimens of the Wiluna fall are minute.

59

Plate 9 (*top*) *The Mundrabilla meteorite, Western Australia, weighing more than 11 tons, shows the broken surface and the flattened striated cone form of the oriented mass. The ablation striations are well seen in this close-up of the mass: note the very slight penetration into the limestone surface; (below) The smaller mass, found 200 metres away. Like the larger it rests on a thick wad of iron shale (a product of weathering by the agency of groundwater). This mass is inverted with respect to the larger mass, having tumbled over after separation, prior to landing. The convex side shown joined onto the concave broken side of the larger mass*

The term micrometeorite has been used variously by different authors: the term can be applied to any meteoritic material of microscopic size. There is no doubt, considering the dust trails left by bolides, that the Earth must accrue a considerable amount of such material from meteoritic sources, though it must be very widely dissipated and difficult to detect in soil or dust samples, even though it can be detected in certain favourable locations—in deep-sea deposits or on polar ice caps, and in samples taken from the upper atmosphere. There is also a form of cosmic dust, not related to ablationary processes, but formed of minute particles from Space which enter the Earth's atmosphere without alteration. Estimates of the amount of ablationary material accrued in dust form by the Earth annually vary: it may be as much as a million tons: certainly it is a very large amount.

Meteorite masses may have a wide variety of shapes: most common are faceted shapes, ablation modifications of fracture surfaces. It is quite certain that most meteoroids orbit in Space in the form of such faceted masses, with the appearance of a fresh broken block of stone, stony iron, or metal. Of course, they are modified by the loss during ablation, and much work has been done on irons, using cosmic-ray-induced isotopes (^{3}He) (Chapter 17), the results of which suggest that there is a wide variation in the degree of ablation loss, from more than 60 per cent in the case of a small mass to about 30 per cent in the case of a moderately large one. Obviously several factors contribute: cosmic velocity, trajectory, size and initial form of mass, and composition of the mass, and we must expect a wide variation in degree of ablation loss.

The ablation of meteorites produces *fusion crust* or *melt skin*, a thin layer, a fraction of a millimetre thick, which envelops the entire mass that falls to the ground. In very rare cases stony meteorites have been known to fall with no fusion crust developed, or a very incomplete fusion crust coating (Krinov cites the Stavropol, USSR, a hypersthene chondrite) and some members of the Kunashak shower (USSR), of similar meteorite stones. Such surfaces are rounded and may show small island patches of crust, centred on projecting metal specks. Very late break up during entry may also result in some surfaces free of crust in fragments arranged in a dispersion pattern, or surfaces with a very thin discontinuous coating of fusion crust (eg, Wiluna, Western Australia—a bronzite chondrite).

Fusion crust is most noticeable on stony meteorites: in development of fusion crust the howardite and nakhlite achondrites are preeminent, their light interior material commonly contrasting with a black crust which shines like black lacquer or patent leather. The common chondrites also usually display black fusion crusts, which,

61

though normally quite matt rather than shiny, contrast strongly with the interior material. Some meteorites that carry pure enstatite (eg, aubrites, enstatite achondrites) may show white fusion crusts, and the crust may even be transparent, showing the crystalline structure through a transparent glassy skin, as in the case of Pesyanoe (USSR). Obviously, the lack of iron in such meteorites has a bearing on the light colour of the crust.

Long sojourn in the soil converts the fusion crust to a brown, iron-stained material and finally removes it altogether, or else circulating groundwater may convert the crust of the buried portion into a white caliche encrustation.

Iron meteorites show a black to bluish crust immediately after fall, but the bluish tint is rapidly lost: exposure, either to the atmosphere or chemical agents following soil burial, alters the appearance of the crust. Surfaces exposed to the air go brown with oxidation, but the oxidation does not penetrate deeply in arid climates, and one may still find fresh metal immediately beneath the very thin oxidised film. In contrast buried material, even in arid climates, tends to become converted to iron shale, which may form a thick pad beneath the

Plate 10 The Haig meteorite, an octahedrite iron showing a perfectly preserved regmaglypt pattern reflecting atmospheric ablation

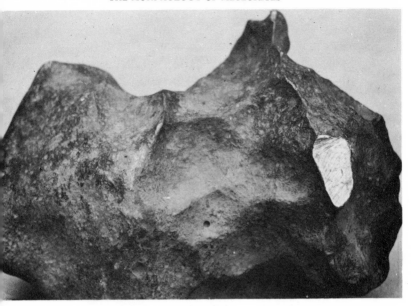

Plate 11 The Kumerina iron, an octahedrite from Western Australia, showing the bright metallic interior, before polishing and etching; also shallow, not well-defined regmaglypts

meteorite. In damp climates, iron meteorites must decay very rapidly, and this is accentuated in areas close to the sea, where the salt carried inland in the sea breezes hastens the process. Meteorite finds are, in fact, very unlikely to be made close to the sea shore, in any climate, though there are exceptions; large masses in particular may survive for a considerable period even under such adverse conditions.

The fusion crust is not obvious in the case of irons though the ablation surface may, of course, be detected by ablation markings—striations and regmaglypts (thumb prints). The former mark the flow of ablating melt from front to rear surface: the latter are round, almond shaped, elliptical or polygonal pits, and grooves, milli-metres or even centimetres in diameter. Iron meteorites tend to show wavy and reticulated regmaglypt patterns covering the primary ablation surfaces in a regular manner. They may also show large cir-cular pits, particularly on rearward facing surfaces in ablation flight: these represent differential ablation of large troilite nodules, which

63

are commonly spherical. In an extreme case these may develop into deep, extensive, irregular cavities as in the case of the Willamette meteorite.

In polished and etched sections, iron meteorites will show a very thin thermally altered zone at the ablation surfaces, a zone in which the Widmanstätten pattern, Neumann lines, or ataxitic microetch pattern is entirely obliterated. The ablation loss tends to round off meteorites: small individual iron masses may be very much rounded off and ball-like masses have even been recorded amongst the smaller irons. Rare forms of iron meteorites are ring-shape (eg, the Tucson, USA, iron) and horseshoe shape (the Mt Magnet iron, Western Australia).

The stones may have the form of oriented cones, but more often show a faceted form or modifications of faceted form. The fusion

Plate 12 The Milly Milly iron from Western Australia, showing a cut face after etching: the characteristic etch pattern (Widmanstätten) of an octahedrite is apparent. Note also the deep regmaglypts on the ablated surface

crust composition varies with the meteorite composition; it is, in most cases, largely composed of magnetite (Fe_3O_4), together with a glass broadly of the composition of the meteorite. The crust generally

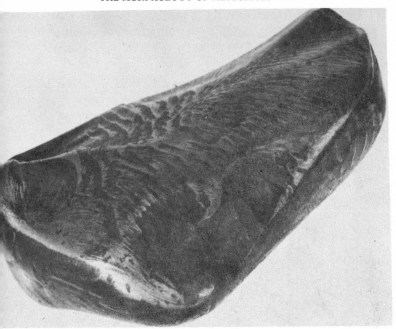

Plate 13 The Nedegolla, India, iron, showing an unusual wave-like ablation pattern on a shiny metallic surface. The meteorite is a heat-modified hexahedrite or nickel-poor ataxite of rare type

has a complex layered structure, consisting of three distinct layers. Stones are very poor conductors of heat, and this fact explains the thinness of the fusion crust, in spite of the very high temperatures involved in ablation.

Regmaglypts are more scattered on stones, though they may form systems of sub-parallel or radiating grooves, and there may be a network of wavy patterns, particularly on the rearward-facing surface in ablation flight: they also tend to be concentrated near the corner on the anterior surface of conical, oriented meteorites, where the flow lines curve over the abrupt corner onto the rearward-facing face. Anterior-facing surfaces tend to be smooth and rounded, particularly near the stagnation point, the extreme front point from which the ablation striations and regmaglypt groovings radiate. The fusion crust is very thin and smooth on the anterior surface, and may appear a lighter colour, possibly brownish-tinged; whereas on the rearward-facing surface it is rough, black, and thickened.

65

The exterior characteristics of fusion crusts have been classified by Krinov. Complex spattering effects appear to be involved in the development of some of the more complicated type surfaces:

1 *Frontal surfaces*

    Close textured: Smooth and commonly little disturbed by regmaglypts, though marked by striation which may show no relief, being only colour streakings. Striations radiate from the stagnation point. Characteristic of irons.

Plate 14 The Middlesbrough, England, stone, an olivine hypersthene chondrite, showing the anterior surface in ablation flight, with regmaglypt patterns streaming radially out and backwards from the stagnation point—a perfect example of oriented form

Knobby: Fine, angular knobs complicate an otherwise smooth surface. In stones, the knobs are found to coincide with lumps of metal.

2 *Lateral surfaces*

Striated: Thin striations on a close-textured crust mark the flow across the lateral surface, commonly having relief (ie, being ridge and groove striations). The striations tend to curve and even to change direction abruptly. Sometimes, thin striae form a system superimposed on an earlier wide, flat system of striae.

Ribbed: Only found on stony meteorites, a form of underdeveloped striation.

Plate 15 The Woolgorong stone, an olivine-hypersthene chondrite that fell in Western Australia in 1960; a perfect example of oriented form, showing strongly contrasted anterior (right) and posterior (left) ablation surfaces. Ablation striae, regmaglypts, and the rougher, thickened posterior ablation surface fusion crust are all apparent

Net: Typical of the more friable stony meteorites, especially near protuberances, the striae form a fine network like stitchwork.

Porous: Typical of oriented stones and irons in which the coin is not abrupt, and of depressions in the surface, it is a microscopic porosity of the crust.

3 *Rear surfaces*

Warty: Common on iron meteorites, warty protuberances may form a texture visible to the naked eye in the case of large masses, or a microscopic texture in the case of smaller masses. It is rarely seen on stony meteorite surfaces, and if it is present is poorly developed within depressions or at the edges of rearward-facing surfaces.

Scoriaceous: Characteristic of stony meteorites, but rare on surfaces of irons, and then found only near sharp edges and protuberances. The appearance is that of a clinkered slag.

Krinov concludes that close textured and knobby crusts reflect high atmospheric pressure; curvature of lines reflects turbulent flow and complex systems of striae superimposed reflect change of position during ablation—that is aerodynamic instability. The most important conclusion is, however, that the coarseness of striae, nodules,

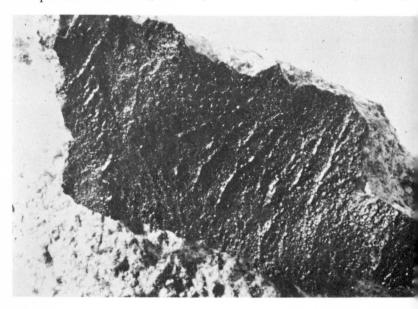

Plate 16 Warty fusion crust on the Barwell meteorite which fell in Leicestershire in 1965

and regmaglypts gives an indication of the size of the original mass. In some cases, such as the Sikhot-Alin meteorites, a set of small secondary networks related to the geometry of the mass fragmented off at 6 km above the Earth's surface, may be seen, with traces also of a coarse regmaglypt pattern unrelated to this geometry, but related to the original large mass before violent fragmentation. Similar effects are seen on masses of stony material of the Wiluna shower. Warty surfaces are due to shedding off of particles of glass into the dust trail, and then a sucking back process as the cosmic velocity of the mass is entirely lost, joining up once again to form a more or less uniform accretionary layer on the still plastic fusion crust surface. Scoriaceous crust is due to internal frothing at the edges and rear surface.

Fusion crusts of stony meteorites commonly show minute polygon tension cracks. Concentric stains, or peeling shell-like patches, are also common, and are related to detachment of single chondrules and development of a pattern centred on the hollow left behind. Specks of metal tend to stand out from fusion crusts of stony meteorites, because metal is a better conductor of heat, and, therefore, melts more slowly, while the stony material, a poor conductor, heats up intensely in a very restricted surface layer. In some cases, minute crystals of olivine (etc) are formed as the fusion crust congeals, giving the crust the appearance of porphyritic basalt.

Plate 17 (see opposite) Wiluna, Western Australia, shower, 1967. (top) Mass with two generations of fusion crust, one rough and less complete, showing mottled: the earlier formed crust showing dark and smooth; (centre) Lump of nickel-iron showing through the first-formed fusion crust. White specks in the crust appear to be regenerated olivine crystallites; (below) Fragment showing only a thin, transparent partial fusion crust, through which the interior of the meteorite looms. This could be a third generation crust of a poor formation of second generation crust. The original mass suffered at least two and possibly three explosive disruptions while in ablation flight

6: The Classification of Meteorites

The first classification of meteorites was the outcome of studies by G. Rose (1863), G. Tschermak (1883), and A. Brezina (1904). The final outcome of the latter author's work was a somewhat unwieldy, structureless classification of the stones, and a rather more satisfactory division of the irons and stony irons into groups and classes. This classification is given in full by E. L. Krinov (1960, p 398) and B. Mason (1962, p 47) but it is obsolescent as far as stony meteorites are concerned, though the numerous symbols are still used in catalogues.

G. T. Prior (1920) introduced the first systematic division of stony meteorites when he replaced the superficial criteria of colour, brecciation, etc, used by Brezina by more meaningful mineralogical and chemical criteria. He recognised that the chemical variations in the metal and silicate components are sympathetic, relating to the oxidation states: and that they seem to obey certain simple rules, which are now commonly referred to as 'Prior's rules'.

1 Total nickel-iron percentage in the chondrites decreases as the nickel-iron ratio in the metal fraction increases.

2 Total nickel-iron percentage in the chondrites decreases as the iron/magnesia ratio in the silicate fraction increases.

From this pattern of variation four meteorite groups were derived —groups which embrace all meteorites, not simply the chondritic stones (Table 2).

73

TABLE 2

Prior's Classification (Modified by G. J. H. McCall)

Class	Type 1	Type 2	Type 3	Type 4
Nickeliferous Iron Magnesium Silicates	FeNi Ratio 13 Enstatite, Clinoenstatite	8–13 Bronzite, Clinobronzite, Olivine	2–8 Hypersthene, Clinohypersthene, Olivine	Pyroxenes (mainly Clino-), Olivine
Feldspar	Oligoclase	Oligoclase	Oligoclase	Anorthite
Irons (mainly Nickel Iron)	(Ataxites),* Hexahedrites, Very coarse and coarse Octahedrites	Medium to fine Octahedrites	Very fine Octahedrites, Ataxites	Oktibbeha County– Extreme nickel-rich ataxite
Stony-Irons (Nickel Iron and Silicates about equal)	†Enstatite Olivine Stony Irons, (Mt Egerton)	Most Pallasites Siderophyre Lodranite Mesosiderites	A few Pallasites	
Stones 1 Chondrites	Enstatite Chondrites % NiFe 20–28	Olivine Bronzite Chrondrites % NiFe 16–21	Olivine– Hypershene Chondrites % NiFe 1–10	
2 Achondrites	Aubrites (Enstatite Achondrites)	Ureilites (Olivine Pigeonite achondrites)	Diogenites, Chassignite (Hypersthene and olivine achondrites)	Angrite, Nakhlites, Eucrites, Howardites

* Thermally modified hexahedrites ('nickel-poor ataxites').
† These additional types would fit quite easily into Prior's classification.

TABLE 3

The Prior–Mason Classification (Modified by G. J. H. McCall)

STONES (Aerolites)

(i) Chondrites. Classified on basis of mineralogy (reflecting decrease in free iron, increase in oxidised iron) and possession of chondrules
  (a) Enstatite chondrites
  (b) Olivine bronzite chondrites
  (c) Olivine hypersthene chondrites
  (d) Olivine pigeonite chondrites (carbonaceous chondrites type III)
  (e) Carbonaceous chondrites (types I and II)

(ii) Achondrites. Classified on basis of mineralogy and calcium content

    (a) Calcium-poor[1]
      1 Enstatite achondrites (Aubrites)[2,4]
      2 Hypersthene achondrites (Diogenites)[3,4]
      3 Olivine achondrite (Chassignite)
      4 Olivine pigeonite achondrites (Ureilites)
    (b) Calcium-rich
      1 Augite achondrite (Angrite)
      2 Diopside olivine achondrites (Nakhlites)
      3 Pyroxene-plagioclase achondrites
        (*a*) Hypersthene-bytownite (Howardites)
        (*b*) Pigeonite-bytownite (Eucrites)[9]

STONY-IRONS (Siderolites).[11] Classified on basis of mineralogy and chemistry

    (a) Olivine stony-irons (Pallasite)
    (b) Bronzite tridymite stony-iron (Siderophyre)[5]
    (c) Bronzite olivine stony-iron (Lodranite)
    (d) Pyroxene plagioclase stony-irons (Mesosiderites)*[6,7,8,10]

IRONS (Siderites). Classified by structure of alloy revealed by etching with acid (which reflects the nickel/iron ratio)

    (a) Hexahedrites[13,16,17] Ni 4–6%
    (b) Octahedrites[14] Ni 6–14%
      1 Width of Kamacite lamellae[12] in Widmanstätten figure $>2{\cdot}0$ mm. Coarse 6–8% Ni
      2 Width of Kamacite lamellae $0{\cdot}5$–$2{\cdot}0$ mm. Medium 7–9% Ni
      3 Width of Kamacite lamellae $<0{\cdot}5$ mm. Fine 8–14% Ni
    (c) Nickel-rich ataxites,[15] Ni $> 10\%$ (mostly $>14\%$)

Notes to Table 3

1 Amphoterites ('hypersthene olivine achondrites') have been shown to be sub-class of recrystallised chondrites possessed of unusually iron-rich olivine (Mason and Wiik, 1964), and are now included within the olivine-hypersthene chondrites.

2 Bustite is synonymous with aubrite.

3 Rodite is synonymous with diogenite.

4 Chladnite is a term used to cover aubrites and diogenites (embracing all the orthopyroxene achondrites).

5 The sole Siderophyre, Steinbach, is recorded as having olivine present by Tschermak (1885), in addition to bronzite and tridymite.

6 The Mesosiderite Mount Padbury contains tridymite, besides olivine which is a minor constituent of several mesosiderites and a major one of others.

7 Some atypical meteorites such as Bencubbin, Weatherford, and Patwar are commonly classed as mesosiderites—yet the remainder of the group shows a pattern of chemical, textural, and mineralogical unity (McCall, 1965*, 1968), a pattern into which these can in no manner be reasonably fitted. They seem to be enstatite-olivine stony irons, and to require a fifth subdivision.

8 Mount Egerton, an enstatite-bearing achondritic meteorite of stony-iron composition (McCall, 1965†), is quite unlike Bencubbin, Weatherford, and Patwar, and seems to be an unusually metal-rich enstatite-achondrite, displaying the anomalous pseudo-octahedrite etch pattern due to the presence of the *perryite* phase carrying nickel, iron, and silicon first recognised in the unique Horse Creek iron.

9 *Shergottite* denotes a shocked eucrite in which the plagioclase, labradorite, has been converted to a glass, *maskelynite*.

10 *Grahamite* was used originally for stony-irons of the character of mesosiderites which display plagioclase in addition to pyroxene and olivine. The term is obsolete.

11 There is a separation of the stony irons into *Siderolites* (transitions from stone to iron, with nickel-iron forming enclaves in sections—includes Mesosiderites, and the Lodranite): and *Lithosiderites*, similar transitions but with nickel-iron continuous in section. This is unrealistic and obsolescent: some mesosiderites show continuous nickel-iron in section, for example. It is in the nature of a needless and confusing complication—all stony irons should be considered as siderolites, even though it is recognised that, while most possess continuous metallic networks enclosing silicates, the metal areas in some may be very patchy and discontinuous, even nodular.

12 The metallography of nickel-iron alloys is treated fully and admirably by Perry (1944); though, of course, his is by no means the last word as far as interpretation goes.

13 Nickel-poor ataxites are recorded, but are now considered to represent heated zones in hexahedrites.

14 Ataxitic meteorites of the composition of octahedrites are, in some cases, heated zones in octahedrites (*Metabollites*).

15 The term *ataxite* is a poor one—few ataxites are 'structureless'—only those with more than 25 per cent nickel show little apparent

---

* Paper on the Dalgaranga crater, Bibliography Chapter 4.

† Paper on Mt Egerton meteorite, Bibliography Chapter 10.

structure whereas the others reveal fine etch patterns, composed of plessite inset with kamacite inclusions fringed by taenite, but do not show the lamellar intersections that are characteristic of the Widmanstätten figures of octahedrites, nor the Neumann lines which provide the only etch patterns of hexahedrites.

16 Very coarse octahedrites may show no lamellar pattern, but only an irregular, coarse granularity. There are some meteorites which show areas of this character and other areas of hexahedrite character.

17 The Horse Creek *pseudo-octahedrite* and Mount Egerton meteorite contain an iron-nickel alloy including silicon which might be regarded as requiring a fourth subdivision of meteoric iron (see Note 8 above). However, they contain about 6 per cent nickel in the alloy which is close to the composition of hexahedrites, with which they are, perhaps, best classified, as anomalous hexahedrite types.

A later classification has been derived by Van Schmus and Wood in 1967 to be applied alongside this classification to chondrites: it includes a number of mineralogical and textural features, essentially parameters of recrystallisation and equilibration state. This classification is given in the later chapter dealing with recrystallisation. There are also certain geochemical classifications (for example L and H chondrites of Urey and Craig and cooling rate classifications of Goldstein and others associated with gallium/germanium contents): these again are discussed in later chapters, being of specialised application.

7: The Ratio of Falls to Finds of Various Types of Meteorites—Geographical Statistics

Hey, in his second (1966)* revision of Prior's catalogue, lists 661 iron meteorites and 1,199 stones, yet only 43 certain witnessed falls of iron meteorites are recorded, as against 723 certain witnessed falls of stones! Approximately two-thirds of all stony meteorite occurrences known are derived from observed falls. The inference is obvious, that stones disintegrate very rapidly under the influence of terrestrial weathering agencies (rain, wind, chemical solutions moving over and through the ground surface, and into the meteorite mass). Thus the ratio of 96 per cent falls of stones to 4 per cent of irons is considered to represent a true proportion arriving into the Earth's atmosphere from Space, in spite of the scanty recovery of stones from chance finds except in extremely arid countries with hard-pan surfaces such as Western Australia. Yet this ratio must *not* be taken as representing the mass or volume proportions in Space, for iron meteorite masses are more dense and in general much larger than stony meteorite masses (suggesting less ready disintegration in Space and during ablation flight) and statistics of *numbers* of occurrences do not give a proper evaluation of the volume, or mass of material of each kind orbiting in Space: such proportional evaluations are, in fact, very difficult to derive accurately by any method. The 24:1 stone:iron ratio derived from fall statistics is certainly a much higher ratio than either the mass or volume ratios, which must be considerably lower. The latest meteorite statistics, covering geographical distribution, falls and finds of the classificatory types, are given in Table 4.

The uneven geographical distribution of meteorites has been considered by most earlier workers to reflect such terrestrial factors as climate, presence or absence of soil, and vegetation density in relation to uneven find statistics (and even the acts of Man in preparing weapons from irons): the distribution of fall observations, also

* Bibliography, Chapter 5.

remarkably uneven statistically, was likewise considered largely to reflect population density. However, modern researches show that the same rare type of meteorite (diogenite) has fallen twice within a decade in central India, and that two ureilites, of extreme rarity and of unlike character and weathering state, have been found within 20 miles of one another in Western Australia.

As we gather in more and more distributional statistics (this was a very neglected field of meteoritics, until recently), it becomes clear that the Earth has been preferentially, not evenly, peppered by meteorites from Space. The same types occur in clusters, geographically: for example, hexahedrites in South and Central America: the large irons in Mexico and South-West Africa. It becomes clear that the observation of Barcena in 1876 that 'there is a peculiar property, difficult of explanation, which the Mexican soil has, in attracting meteoritic material' was an early intimation of the truth now emerging: that orbital groupings control terrestrial distributions to a considerable extent, and that successive interferences between meteorite orbits and the Earth years apart can well bring down meteorites of the same type at geographically close points. In the same context, the present author has remarked on the strange lack of large meteorites on the Nullarbor Plain, Western Australia, an area of optimum recovery potential because of preservation and lack of burial.

TABLE 4

List of Meteorites—Falls and Finds, Totals—by Types
(After Hey, 1966)
(Italic figures indicate doubtful occurrences)

	Falls		Finds		Total
IRONS					
Total	43	*18*	590	*16*	667
Nickel-rich ataxites	1		41		42
Octahedrites	30	*2*	439	*3*	474
Off	—		22		22
Of	4		63		67
Om	14		203	*1*	218
Og	4		92	*2*	98
Ogg	2		22		24
Unclassified octahedrites	6	*2*	31		39
Brecciated	2		5		7

THE RATIO OF FALLS TO FINDS OF VARIOUS TYPES OF METEORITES

TABLE 4—(contd.)

	Falls		Finds		Total
Metabolites (thermally altered)	—		3		3
Hexahedrites	6		48	*1*	55
'Nickel-poor ataxites' (thermally altered hexahedrites)	1	*1*	26		28
Unclassified irons	3	*15*	30	*12*	60
STONY-IRONS					
Total	12		59	*3*	74
Pallasites	4		40	*2*	46
Siderophyre	—		1		1
Lodranite	1		—		1
Mesosiderites	6		16		22
'Enstatite olivine' stony-irons	1		1		2
'Enstatite stony-iron'	—		1*		1*
Unclassified	—		—	*1*	1
STONES					
Total	724	*42*	428	*20*	*1,214*
Chondrites	625	*5*	404	*10*	1,044
CEn	11		6		17
CBr	225	*1*	190	*4*	420
CHy	313	*2*	192	*3*	510
(Amphoterites)	39		10		46
CPig CCIII	15		—		15
CC I and II	20	*1*	—		21
Unclassified	41	*1*	16	*3*	61

* Mt Egerton probably better regarded as an *aubrite* with an unusual amount of metal of 'Pseudo-octahedrite' character.

	Falls		Finds		Total
ACHONDRITES					
Total	59	*1*	10		70
Calcium-poor division					
Aubrites (enstatite achondrites)	7		2		9
Diogenites (hypersthene achondrites)	8		—		8
Chassignite (olivine achondrite)	1		—		1
Ureilites (olivine pigeonite achondrites)	2		3		5
Calcium-rich division					
Nakhlites (diopside olivine achondrites)	1		1		2
Angrite (augite achondrite)	1		—		1
Eucrites (pigeonite plagioclase achondrites)	26	*1*	3		30
Howardites (plagioclase hypersthene achondrites)	12		1		13
Unclassified achondrites	1		1		2
Unclassified stones	39	*35*	15	*8*	97
Unclassified meteorites	7	*66*	2	*12*	87
Strange objects reported as meteorites					
Copper meteorites	—	*1*	—		1
Limestone meteorites ('Calcarite')	—	*1*	—		1
Sandstone meteorites	—		—	*2*	2
Grand total (all meteorites recorded)	786	*127*	1,080	*51*	2,044

Geographical Statistics (After Hey, 1966)

	Falls		Finds		Total
Europe	315		64	*12*	476
Asia	214	*21*	72	*12*	319
Africa (+ Indian Ocean)	78	*5*	40	*2*	125
North America	129	*8*	659	*13*	809
South America	37	*6*	87	*3*	133
Australasia	13		130	*7*	150
Antarctica	—		4		4
(Elsewhere: eg, Atlantic Ocean, not located)	—	*2*	1	*2*	5

THE RATIO OF FALLS TO FINDS OF VARIOUS TYPES OF METEORITES

Table 4—(*contd.*)

	Falls		Finds		Total
COUNTRIES					
Norway	7		2		9
Sweden	9	*3*	3	*1*	16
Finland	5		5		10
Denmark	3	*1*	—		4
Scotland	3	*2*	—	*1*	6
Ireland	6		—		6
Isle of Man	—	*1*	—		1
Wales	2	*1*	—		3
England	10	*10*	—		20
Holland	3	*2*	—		5
Belgium	4	*3*	—		7
Germany	27	*20*	10		57
Poland	8		5	*2*	15
France	56	*6*	3		65
Switzerland	3	*3*	—		6
Austria	3	*1*	1	*1*	6
Czechoslovakia	15		8		23
Hungary	6	*3*	1	*1*	11
Rumania	6		1		7
Portugal	4	*1*	2		7
Spain	22	*5*	3	*1*	31
Italy	26	*17*	—		43
Yugoslavia	9		2		11
Bulgaria	4		—		4
Turkey	6	*3*	1	*1*	11
Greece	1	*3*	—	*2*	6
USSR	92	*3*	46	*4*	145
Outer Mongolia	3		3		6
Syria	2		—	*1*	3
Lebanon	—		—	*1*	1
Jordan	1		—		1
Arabia	3		17	*4*	24
Iraq	1	*1*	1		3
Iran	1	*1*	—		2
COUNTRIES					
Afghanistan	1	*1*	—		2
Tibet	1		—		1
Pakistan	21	*1*	—		22
India	102	*7*	6		115
Ceylon	1		—		1
Cambodia	2		—		2
Burma	3		—		3
Vietnam (S)	3		—		3
China	7	*1*	3		11
Korea	2		1		3
Japan	25	*5*	9	*4*	43
Java	7	*1*	1		9

83

	Falls		Finds		Total
Philippine Is	3		1		4
Morocco	1		1		2
Algeria	6		5		11
Tunisia	1		—		1
Lybia	—		1		1
Egypt	2	*1*	3	*1*	7
Mauritania	—		1		1
Mali	1	*1*	1		3
Niger	3		—		3
Chad	1		—		1
Volta	3		—		3
Nigeria	8		3		11
Cameroon	2		—		2
Central African Republic	1		—		1
Congo	5		—		5
Sudan	4		—		4
Ethiopia	3		—		3
Somalia	2		1		3
Uganda	1		—		1
Kenya	3		—		3
Tanzania	7		1		8
COUNTRIES					
Angola	1		1		2
Rhodesia	1		—		1
Zambia	1		—		1
Malawi	1		1		2
South West Africa	1		4		5
Bechuanaland	—	*1*	—		1
Union of South Africa	17	*1*	17		35
Mauritius	1		—		1
Greenland	—		2		2
Canada	9		23		32
USA	109	*8*	52	*11*	710
Mexico	10		48	*1*	59
Guatemala	1		—		1
Honduras	1	*1*	—		2
Costa Rica	1		—		1
Cuba	—		1		1
Jamaica	—		1		1
Colombia	—		1		1
Peru	—		2		2
Bolivia	—		3		3
Brazil	17	*4*	13	*2*	36
Paraguay	1	*1*	—		2
Argentina	18	*1*	23	*1*	43
Chile	1		45		46
New Guinea	1		—		1

THE RATIO OF FALLS TO FINDS OF VARIOUS TYPES OF METEORITES

TABLE 4—(contd.)

	Falls		Finds		Total
New Caledonia	1		—		1
Australia	9	*1*	126	*7*	143
Western Australia	1	*1*	46	*2*	50
Northern Territory	11		—		11
Queensland	2		4	*3*	9
South Australia	1		23	*1*	25
Victoria	—		6		6
New South Wales	4		33		37
Tasmania	1		3		4
New Zealand	1		4		5

8: The Mineralogy of Meteorites

The mineralogy of meteorites has been covered at length by Mason in a special treatment published in 1963 and in several later papers—including an important paper published in 1967. This chapter is a summary largely based on his detailed accounts (where no reference is given to an authority named, see these publications). Meteorites do not contain *elements* not known to us on Earth in natural occurrence, though such elements may have previously existed and been lost by radioactive decay (Chapter 17). However, while some of the minerals are familiar to us in terrestrial igneous rocks—for example, olivine, pigeonite, augite, enstatite, bronzite, diopside, plagioclase, chromite, tridymite, quartz, and cristobalite; other minerals have not so far been recognised in terrestrial rocks—for example schreibersite, cohenite, osbornite, and oldhamite. These, if recognised, virtually establish the meteoritic nature of the material. The alpha and gamma forms of nickel-iron, and the sulphide mineral troilite, the major components of meteoritic irons, are not of exclusive meteoritic occurrence, though extremely rare in terrestrial occurrence.

Some mineral species found in meteorites form *isomorphous series*: that is, they form a continuous chemical gradation from one end member to another, a gradation in which there is progressive substitution of one end member for another. In such series, the crystallographic form remains essentially the same, despite slight variations in the lattice geometry of the crystal. Such changes are detectable by the use of X-ray diffractometry. They are due to substitution of ions of different size, which produce slight variations in the crystal geometry without varying the essential crystallographic form. Such isomorphous series are denoted by reference to the two end members, which must have the same basic chemical formula—for example Fo_{21} means an olivine with 21 per cent forsterite and 79 per cent fayalite. Some meteorites contain minerals that represent a part of such series

87

that is never represented in terrestrial rocks—for example the enstatite variety Fs_0 which has none of the iron-rich end member

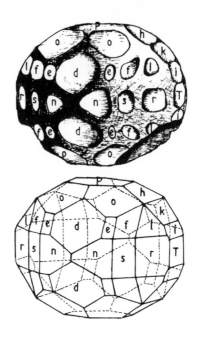

Fig 9 Olivine grain from the pallasite Krasnojarsk, showing faceting, an expression of the natural crystal faces. The sketch below it shows the complete faces represented by the facets, mutually intersecting (after Koksharov)

ferrosilite, and is thus entirely composed of the magnesium-rich enstatite end member $MgSiO_3$. Likewise troilite, the end member of the terrestrial pyrrhotite series, is mainly restricted to meteorites. Many of the isomorphous series are thus familiar in terrestrial igneous rocks and ores, though the actual members of the series found in meteorites may be unique to or virtually restricted to meteorites.

Meteoritic material does display other disparities to terrestrial rock material, in addition to those that may be classed as mineralogical: for example, in isotopic ratios for the various elements, in trace element abundances, and in inert gas contents. These peculiarities are the subject of an extensive field of geochemical/geophysical study that is telling us much about the origins of the Earth and Solar System.

The mineral grains in meteorites are commonly well formed. The classic drawing by Koksharov of the crystal faces of an olivine in the Krasnojarsk, Siberia, pallasite, illustrates this. Such a proliferation of faces is seldom, if ever, attained in olivines from terrestrial igneous rocks, and suggests slow and unimpeded crystallisation. In contrast, many grains in chondritic meteorites are quite anhedral, that is without crystal form, and areas of glass and incipient crystallites occur, as in volcanic rocks on Earth. Such occurrences indicate rapid chilling, in contrast to the Krasnojarsk olivine.

The relative distribution of the principal minerals in meteorites is summarised in Table 5.

TABLE 5

Modal Percentage of the Various Major Mineral Phases in Meteorites (By Volume) (After Wood, 1963)

| Minerals | Irons | Classes of Meteorite | | | |
		Pallasites	Mesosiderites	Achondrites	Chondrites
Nickel-iron	98·34	50·0	45·0	1·57	10·58
Olivine	—	48·0	1·5	12·82	42·31
Pyroxenes	—	—	30·6	62·25	28·91
Feldspars	—	—	16·4	20·75	11·82
Troilite	0·12	0·3	3·1	1·53	5·01
Schreibersite	1·12	0·2	2·6	0·40	—
Chromite	—	—	0·8	0·68	0·78
Cohenite	0·42	1·5	—	—	—
Apatite	—	—	—	—	0·67

The minerals so far recognised in meteorites are enumerated in Table 6. This list is reasonably comprehensive, though a few rare and less important species are omitted. It will be noted that the minerals are all anhydrous, with the exception of a few minerals found in the carbonaceous chondrites of types 1 and 2. The common hydrous rock-forming minerals such as biotite and the amphibole, hornblende, are not found at all.

TABLE 6

Minerals Occurring in Meteorites

Mineral group	Name	Formula	Occurrence in Meteorites
Native metals	Nickel-iron kamacite taenite plessite (para-eutectoid)	Fe(Ni)	Occurs in virtually all meteorites, though rare in achondrites and carbonaceous chondrites. Nickel ranges from 5·5% to 66%
	Copper	Cu	Specks occur in irons and also in chondrites. There is a supposed meteorite (reported in the possession of H. H. Nininger) from Eaton, Colorado, entirely composed of a copper alloy. Its validity is regarded as questionable by most authorities. There is also a seventeenth-century report of copper meteorites in the form of a shower of highly cupriferous stones at Ermendorf, Saxony, but this material is lost.
	Gold	Au	Free gold is reported from Wedderburn, Victoria, by A. B. Edwards: otherwise gold is only known in trace amounts.
Native non-metals	Diamond (Lonsdaleite) Graphite (Cliftonite)	C	Diamonds are known from ureilite achondrites, and there are reported occurrences in chondrites. The octahedrite material from Cañon Diablo, Arizona, associated with Meteor Crater (Chapter 20), also contains diamonds. The ureilite North Haig contains the hexagonal form lonsdaleite, experimentally synthesised by shock. Cubical aggregates of black graphite (cliftonite) that occur in the Youndegin, Western Australia (Og) and Toluca, Mexico (Om) irons may be paramorphs of diamond. Controversy still exists concerning the question of whether diamonds indicate high pressure crystallisation in the parent body: or a product of shock, either cosmic during asteroidal break-up or on terrestrial impact, the material being originally free carbon. Recent research seems to favour an origin in shock, and there is some reason to believe that both types of shock may have produced diamonds in meteorites.
	Sulphur	S	Free sulphur occurs in type 2 carbonaceous chondrites.

TABLE 6—(*contd.*)

Mineral group	Name	Formula	Occurrence in meteorites
Oxide (spinel)	Magnetite	Fe_3O_4	Primarily crystallised extra-terrestrial magnetite is a component of some carbonaceous chondrites of types 1, 2, and 3 and also of calcium-rich achondrites and mesosiderites. Fusion crusts on meteorites also largely consist of magnetite of secondary origin, due to atmospheric oxidation. With haematite, limonite and goethite, it also results from terrestrial oxidation during surface weathering, being a component of iron shale and shale balls.
Oxide	Rutile	TiO_2	Detected by use of the electron probe in some irons, stony-irons and stones.
	Spinel	$MgAl_2O_4$	Only recognised in carbonaceous chondrites of type 1 and in devitrified glass areas within carbonaceous chondrites of type 3 (Allende, Mexico).
Oxide (silica polymorph)	Quartz	SiO_2	Found in two enstatite chondrites: St Marks, South Africa and Abee, Canada (Dawson *et al.*, 1960).
	Tridymite	SiO_2	Found in calcium-rich achondrites as a late, interstitial phase: also in hypersthene achondrites, mesosiderites, and the unique siderophyre. Is present in the enstatite achondrite from Indarch, USSR. The presence of this mineral phase is supposed to indicate a pressure environment of crystallisation below 30 MN/m^2 on account of the instability of the silica polymorph above this pressure.
	Cristobalite	SiO_2	Only recognised in the enstatite chondrite from Abee, Canada (Dawson *et al.*), though there is a tentative identification by Cohen in one iron, a brecciated hexahedrite from Kendall County, Texas.
Oxynitride	Sinoite	Si_2N_2O	Found in several enstatite chondrites (Anderson, Keil and Mason).
Phosphate	Apatite	$Ca_5(PO_4)_3$	Common in stones, the variety being a chlor-apatite.
	Merrillite	$Na_2Ca_3(PO_4)_2$	Difficult to distinguish from apatite: occurs similarly.
	Farringtonite	$Mg_3(PO_4)_2$	Present in Springwater, Canada, pallasite (Dufresne and Roy).
Carbide	Moissanite	SiC	Doubtful recognition in the Cañon Diablo octahedrite: it is probably introduced carborundum.

TABLE 6—(contd.)

Mineral group	Name	Formula	Occurrence in meteorites
	Cohenite	Fe_3C	Occurs in several irons of type Og including Cañon Diablo and the Youndegin group: it is also found in the Abee, Canada, enstatite chondrite (Dawson, Maxwell, and Parsons). It is supposed to indicate cooling at high pressure (c. 250 MN/m^2) by Ringwood, but Mason questions this conclusion.
Silicide	Perryite	$[Fe(Ni)]_2Si$	This phase has been found in the 'pseudo-octahedrite' Horse Creek, Colorado, and the metal fraction of the unique Mt Egerton, Western Australia, stony-iron (Chapter 10). The metal has the chemical composition of the hexahedrite class of iron, and the strange finely-ruled etch pattern is now regarded as due to this phase by E. P. Henderson, not schreibersite as was at first believed. Silicon also goes into the metal phase of enstatite chondrites (Ringwood, 1961).
Phosphide	Schreibersite (Rhabdite)	$[Fe(Ni)]_3P$	Present in most irons and stony-irons, but usually only in small quantities. It may form shells around troilite nodules. It is particularly associated with the hexahedrites. The variety *rhabdite* occurs as thin plates which are commonly oriented sympathetically to the Widmanstätten pattern.
Nitride	Osbornite	TiN	Found in the Bustee, India, aubrite (Bannister, 1941).
Sulphide	Troilite	FeS	The end member of the familiar pyrrhotite mineral series ($FeS_{(1+n)}$), found in many terrestrial ores etc, this mineral occurs in most meteorites. In irons, it forms rounded or irregular nodules or patches, commonly associated with graphite. It also occurs as rods and plates, which may be sympathetically arranged with the Widmanstätten pattern. It may also occur in the form of small, randomly distributed grains and late veinlet fillings, in chondrites and mesosiderites.
	Oldhamite	CaS	Present in enstatite chondrites and enstatite achondrites.
	Alabandite	MnS	Present in the Abee enstatite chondrite (Dawson *et al.* 1960).

92

TABLE 6—(contd.)

Mineral group	Name	Formula	Occurrence in meteorites
	Pentlandite	$[Fe(Ni)]_9S_8$	Present in two carbonaceous chondrites of type 3: Kaba, Hungary, and Karoonda, South Australia (Sztrokay, 1960 and Mason, 1967): also found in a common chondrite from Isoulane, Algeria (Mason, 1967). Found in type 2 carbonaceous chondrite enclaves in the Holyoke, Colorado, olivine bronzite chondrite (Ramdohr).
	Daubreelite	$FeCrS_4$	This mineral is associated with troilite in many irons, especially hexahedrites, and has been recognised in one enstatite chondrite. It is likely to occur in such highly reduced meteoritic assemblages, and not in oxidised assemblages, where chromite takes its place.
	Chalcopyrrhotite	$(Cu, Fe)S$	Not an uncommon accessory in chondrites (Ramdohr, 1963).
	Vallerite	$Cu_3Fe_4S_7$	Formed by decomposition of chalcopyrrhotite and exsolution in pentlandite (Ramdohr, 1963).
	Pyrite	FeS_2	Intergrown with pentlandite in the Karoonda, South Australia, type 3 carbonaceous chondrite (Ramdohr, 1963).
	Sphalerite	ZnS	Traces reported from enstatite chondrites, Hvittis, Finland, and Pillisfer, Estonia (Ramdohr, 1963).
Chloride	Lawrencite	$FeCl_2$	Present in meteorites of all types, and responsible for a brownish deliquescence on exposure to the atmosphere.
Carbonate	Calcite	$CaCO_3$	Only recorded by L. Kvasha from the Boroskino, USSR, carbonaceous chondrite of type 2.
	Dolomite Magnesite	$CaMg(CO_3)_2$ $MgCO_3$	Only found in carbonaceous chondrites of type 1.
Titanate	Ilmenite	$FeTiO_3$	Occurs in some stones.
Chromate (spinel)	Chromite	$FeCr_2O_4$	Common in stony meteorites. It is an Al and Mg rich variety due to ionic substitution in the lattice.
Sulphates (hydrated)	Gypsum Epsomite Bloedite	$CaSO_4 \cdot 2H_2O$ $MgSO_4 \cdot 7H_2O$ $Na_2Mg(SO_4)_2 \cdot 4H_2O$	Found only in carbonaceous chondrite meteorites of type 1.

93

TABLE 6—(contd.)

Mineral group	Name	Formula	Occurrence in meteorites
Silicates	Olivine (Forsterite, Chrysolite, Hyalosiderite, Hortonolite, Fayalite)	Mg_2SiO_4 (iso-Fe_2SiO_4 morphous series)	Olivine is the most abundant silicate in meteorites: though it is rare in calcium-rich achondrites and enstatite chondrites. It is an undersaturated mineral with respect to silica, and thus few meteorites display silica oversaturation. The magnesian variety, forsterite, occurs in enstatite achondrites and mesosiderites, as well as ureilites and the unique Bencubbin stony-iron (Chapter 10). The olivine of pallasites and common chondrites is the variety chrysolite. The amphoterites and carbonaceous chondrites of type 3 contain the iron-rich variety hyalosiderite or hortonolite.
	Pyroxenes 1 Orthopyroxene (orthorhombic crystal system) Enstatite Bronzite Hypersthene	$MgSiO_3$ $(Mg,Fe)SiO_3$ $FeSiO_3$	Next most abundant of the silicate minerals in meteorites, the orthopyroxenes are represented by enstatite or clinoenstatite pseudomorphing it in enstatite chondrites and achondrites, and in the unique Bencubbin and Mount Egerton stony-irons (Chapter 11): it is also found in type 2 carbonaceous chondrites. Bronzite is more common than hypersthene in meteorites (taking the demarcation at the modern boundary Fs_{30}). It occurs in common chondrites, hypersthene achondrites, howardites, eucrites, amphoterites, the lodranite and the siderophyre.
	2 Clinopyroxene (Monoclinic crystal system) Clinoenstatite Clinohypersthene Pigeonite Augite Diopside	$MgSiO_3$ $(Ca,Mg,Fe)SiO_3$ $(Ca,Mg,Fe)SiO_3$ $(Ca,Mg,Fe)SiO_3$	Clinoenstatite is nowadays regarded as a shock modification of enstatite, and its occurrence is thus the same. The mineral clinohypersthene appears ill defined, but most authorities prefer to term the lamellar clinopyroxene present in many unrecrystallised common chondrites as clinohypersthene, not pigeonite, which is common in achondritic meteorites, eucrites, howardites, and some mesosiderites: also ureilites. It occurs in carbonaceous chondrites of type 3. Augite and diopside have the same general formula as pigeonite, but have different limiting proportions for the three bases, forming part of another isomorphous series which, in terrestrial igneous rocks, also takes in Na. Augite, so common on Earth, occurs only in the unique angrite, and in devitrified glass areas in carbonaceous chondrites of type 3 (Allende, Mexico). Diopside occurs in the two nakhlites, and as

TABLE 6—(*contd.*)

Mineral group	Name	Formula	Occurrence in meteorites
			traces in chondrites, mesosiderites, achondrites, etc, as an exsolution product of bronzite or hypersthene. It is also present in enstatite and hypersthene achondrites, and the sole olivine achondrite as a minor phase.
Feldspars Plagioclase	Anorthite An_{90} Bytownite An_{70} Labradorite An_{50} Andesine An_{30} Oligoclase An_{10} (albite*) * Not found in meteorites.	isomorphous series $CaAl_2Si_2O_8$ ↑ ↓ $NaAl_2Si_3O_8$	Bytownite is the usual plagioclase in howardites, eucrites, nakhlites, and mesosiderites. True anorthite is extremely rare in meteorites, though very commonly recorded in early descriptions.
			Similarly labradorite is rare. Oligoclase is the common feldspar of chondrites, and also occurs in calcium-poor achondrites as a minor component. Oligoclase-andesine is not unknown in chondrites.
	Maskelynite		This is an isotropic mineraloid, a shock product from plagioclase—translucent, like plagioclase in transmitted light in thin section, it is, however, isotropic between crossed nicols (ie, in polarised light). It occurs in the shergottite (Chapter 11), and in recrystallised common chondrites, after oligoclase: in the latter case it may be imperfectly isotropic. The occurrence of oligoclase in calcium-poor achondrites and common chondrites appears anomalous to petrologists for terrestrially it is characteristic of low-temperature igneous and metamorphic assemblages, and not of the ultramafic igneous rocks crystallised at high temperatures and compositionally resembling these meteorite types.
	Serpentines (chlorites)	$Mg_6Si_4O_{10}$	The serpentine minerals are not readily distinguishable from the chlorites: several are reported from carbonaceous chondrites of types 1 and 2, and some have obscure names (for example Murray F mineral). Serpentinised olivine grains and veins of serpentine are known from other olivine-bearing meteorites, but appear to be due to secondary terrestrial weathering effects.
	Gehlenite	$MgSi_2O_7$	This member of the familiar melilite isomorphous series found in terrestrial alkaline igneous rocks is found in devitrified glass within chondrules and other devitrified glass areas in the Allende, Mexico, and Leoville, Kansas, carbonaceous chondrites of type 3.

TABLE 6—*(contd.)*

Mineral group	Name	Formula	Occurrence in meteorites
	Garnet (grossular)	$Ca_3Al_2Si_3O_{12}$	Found in devitrified glass patches of the Allende carbonaceous chondrite, type 3, and Leoville, Kansas, meteorite of the same type.
	Sodalite	$Na_8Al_6Si_6O_{24}Cl_2$	
	Nepheline	$NaAlSiO_4$	
	Perovskite	$CaTiO_3$	Found in devitrified glass patches of the same meteorites.

9: The Iron Meteorites

There are three main subdivisions of the iron meteorites:

		Falls	Finds	Total
1	Hexahedrites	6	49	55
2	Octahedrites	27	364	391
3	Nickel-rich Ataxites	—	—	38

The Hexahedrites

These have a statistical peak composition of 5·5 per cent nickel in the nickel-iron alloy. They consist of large cubic ('hexahedron') crystals of kamacite (alpha nickel-iron), which on etching with the nital re-agent (nitric acid, plus alcohol and glycerine) produce only a pattern

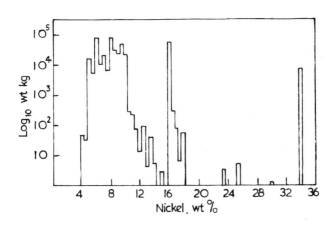

Fig 10 The frequency distribution of recovered weight of iron meteorite material. Note logarithmic scale. By comparison with Figure 11 greater emphasis is given to material at 16 and 34 per cent nickel (after H. J. Axon)

of fine lines called 'Neumann lines'. These lines reflect twinning on the trapezohedron face. There is no other etch pattern evident. Thermal modification effects applied subsequent to crystallisation (due to either a secondary process of metamorphism or recrystallisation within the asteroidal parent body, or else a process that occurred during the approach of the meteoroid at perihelion close to the Sun in its orbital journeying) may give rise to the so called 'nickel-poor ataxites'. These are of hexahedrite composition, but are either devoid of any etch pattern or display partly-obliterated Neumann lines on etching. Uhlig has interpreted Neumann lines as due to violent impact or explosion at some time during the meteorite's *cosmic* history,

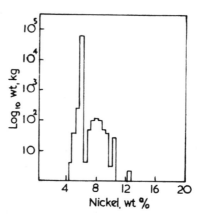

Fig 11 *The frequency distribution of recovered weight of iron meteorites which have been seen to fall. This plot tends to support the suggestion of peaks at 6 and 8 per cent nickel, but, surprisingly, good analyses are not available for all falls of iron meteorite (after H. J. Axon)*

involving a strong mechanical deformation at temperatures probably not exceeding 300°C.

The hexahedrites have an anomalous terrestrial distribution, as Henderson has shown, there being several distinct geographical clusters of masses of identical structure and chemical composition. These include:

   (i) North Chile.
   (ii) Argentina.
   (iii) North Carolina, Georgia, Alabama.
   (iv) Mexico, Texas.

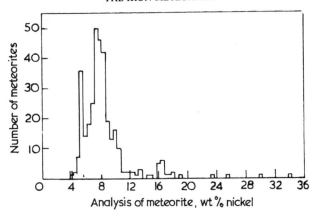

Fig 12 The frequency distribution of the nickel content of iron meteorite analyses. Data taken from M. H. Hey and plotted at ½ per cent nickel intervals by Boustead. This figure is in agreement with the plot at 0·2 per cent nickel intervals which was first made by Yavnel (after H. J. Axon)

It has always been difficult to explain such groupings over wide areas—areas far wider than the maximum limits of elliptical dispersion from a single fall, as noted in Chapter 4: ideas of derivation from more than one of successive orbital pass of masses in Earth capture have lately been put forward, material being supposedly dropped at successive apogee positions, that is, points of closest approach to the Earth, giving the effect of a spray across a region, reflecting the apogee position's variation in relation to the Earth's surface.

These arguments are as yet tentative, and there is some contradictory evidence suggesting that the grouped hexahedrites are not as similar as originally supposed, and thus, may not represent penecontemporaneous arrivals on Earth. The probability seems to be that there is, as suggested, a general orbital control on regional geographic groupings, but that the arrivals represented in such groupings may have been quite widely spaced in terms of years.

The density of hexahedrites is very consistent—c 7.90. The average chemical composition is:

Fe   93·5 per cent ⎫
Ni   5·5 per cent   ⎬ and traces of P, S, Cr, and C.
Co   0·5 per cent ⎭

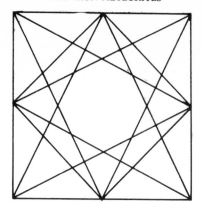

Fig 13 Scheme of systems of Neumann lines, oriented on the surface of a cube (after E. L. Krinov)

Besides kamacite, minute plates of the phosphide mineral schreibersite ('rhabdite') are common, also troilite (sulphide) nodules with associated graphite (or graphite nodules alone, without sulphide).

Gradation from hexahedrite material into the coarsest type of octahedrite material is known, and, indeed, certain falls, such as that of Sikhot-Alin, have included metal of both types. The gradation involves the localised appearance of segregations of the gamma-phase of nickel-iron, taenite, on the margins of the kamacite crystals, and the development of an irregular, coarsely granular texture. The

Plate 18 Neumann lines showing distortion in the rare Gosnells granular brecciated coarse octahedrite from Western Australia. The dark mineral of the round nodule and filling the cracks is probably graphite and troilite, the remainder kamacite. (Polished and etched, cut face)

Plate 19 Granular hexahedrite (or coarsest octahedrite?), Sikhot-Alin, showing Neumann lines

Otumpa irons of Argentina show similar gradations to Sikhot-Alin, and these gradations preclude any theory of special provenance for the hexahedrites. They must come from the same parent bodies as the much more common octahedrites.

The Octahedrites

These display the familiar 'Widmanstätten' figures that the layman commonly associates with meteorites, and form the overwhelmingly predominant class of iron meteorites. The patterns are revealed by etching with nital, picral, or other special acid reagents, but may, on occasion, be visible on weathered surfaces (for example the Tieraco Creek meteorite, Western Australia). The patterns are due to arrangements of broad kamacite (alpha nickel-iron) and narrow taenite (gamma nickel-iron) phases in patterns reflecting a cubic crystallisation. The patterns represent an alloy, comparable with metallurgical alloys which show similar patterns. This alloy is readily comprehensible in terms of experimental studies of nickel-iron melts, though coarse patterns such as these cannot be produced in the laboratory: apparently they require very slow cooling, such as would occur within the interior of a planetary body, for their formation. There have been various diagrammatic representations of the cooling curve applicable to this system: broadly speaking, the whole range of alloys

101

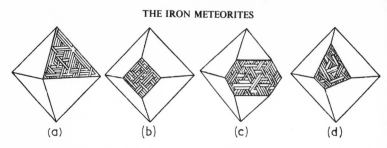

Fig 14 Arrangement of the bars of the Widmanstätten pattern as a function of the section angle: (a) section along octahedral face; (b) section along cube face; (c) section along the face of a rhombic dodecahedron; (d) section in an arbitrary direction (after Tschermak, in 'Lehrb. d. Mineralogie', 1894)

Fig 15 Nickel-iron phase diagram (after Goldstein and Short)

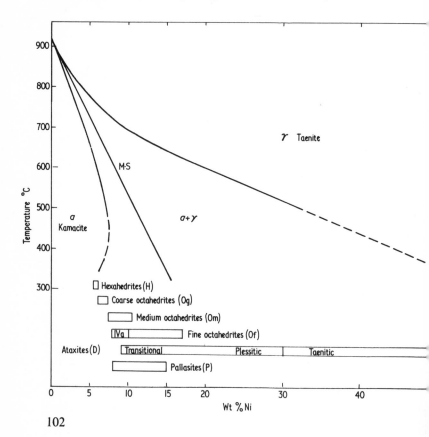

present in nickel-iron of meteorites can be equated, according to Lovering, with fractionation of a parent melt with 11 per cent nickel, modifications of texture being also imposed by variations in cooling rates.

Plate 20 Coarse octahedrite, Ider, Alabama, showing coarse Widmanstätten pattern on cut, etched surface, with a tendency towards localised patches of rounded-off granules (cf the granular hexahedrites, pages 100 and 101). In many of these coarse octahedrites the taenite is only microscopically evident along grain boundaries

Mineral phases (nickel-iron)

$\left\{\begin{array}{l}\text{Kamacite} \quad (\alpha) \\ \text{Taenite} \quad (\gamma)\end{array}\right.$

Plessite   a 'para-eutectoid' consisting of fine intergrowth between the two, often not resolvable into its constituent components without the aid of high power microscopy, and occurring in the form of triangular or polygonal areas or 'fields' interstitial to the kamacite and taenite lamellae.

Octahedrites such as the Mt Edith (No 1) and Avoca meteorites from Western Australia show complex, small-scale intersections of

103

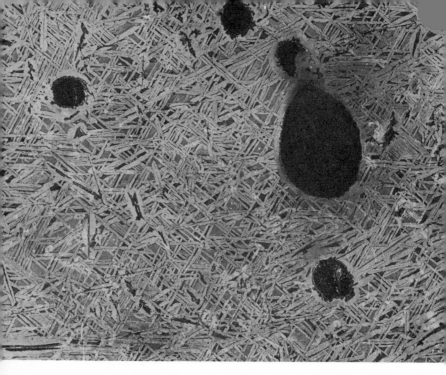

Plate 21 Medium octahedrite, Mt Edith I, from Western Australia, showing Widmanstätten pattern characterised by fine taenite selvages to the kamacite lamellae; fields of the para-eutectoid of kamacite and taenite, plessite, are interstitial to the lamellae and the cut, etched face shows prominent troilite and graphite nodules, walled in by kamacite

lamellae within the fields themselves, as sort of miniature Widmanstätten pattern.

Other mineral phases that may be present in octahedrites are:

Schreibersite	(Fe_3P).
Troilite	(FeS).
Cohenite	(Fe_3C).
Chromite	$(FeCr_2O_4)$.
Graphite	C.

or

Diamond   (many authorities now refer diamonds to secondary shock, either on atmospheric entry or on actual impact, transforming original graphite; and do not accept them as a primary component mineral. This

Plate 22 Medium octahedrite, Avoca, Western Australia, showing a typical Widmanstätten pattern on the cut, etched surface: the taenite forms fine rims to the prominent kamacite lamellae

Plate 23 Polished, etched surface of the Carlton, Texas, fine octahedrite, a typical example, showing kamacite lamellae bordered by very thin taenite lamellae and enclosing fields of plessite. The taenite is only microscopically visible due to thinness of lamellae

may apply to the Cañon Diablo, Arizona occurrence but diamonds in ureilites (p 90) are almost certainly due to earlier shock effects, consequent on

105

collision in Space or events within the meteorite parent body.)

Classification according to band width is given in Table 7.

TABLE 7

Classification of Octahedrites According to Band Width
(After H. J. Axon)

Type	Symbol	Projected width (according to Brezina)	True width (according to Lovering)	Approx. Ni %
Coarsest	Ogg	>2·5 mm		
Coarse	Og	1·5–2·5 mm	>2·0 mm	6·4–7·2
Medium	Om	0·5–1·5 mm	0·5–2·0 mm	7·4–10·3
Fine	Of	0·2–0·5 mm	<0·5 mm	
Finest	Off	<0·2 mm		7·8–12·7
Ataxite	D	No Widmanstätten kamacite		>10·8

The coarsest octahedrites are relatively rare and are of irregular structure, granular in character rather than geometrical: the medium octahedrites are the most common variety and show regular geometrical patterns. The finest octahedrites are relatively rare.

Octahedrites may, like the hexahedrites, display superimposed effects of thermal metamorphism, partly or completely obliterating the etch pattern. Such meteorites have been called ataxites or metabolites: they are, however, quite distinct from the true ataxites, the chemically defined nickel-rich ataxite class.

Nickel-rich Ataxites

This is the third major division of the iron meteorites. The term ataxite is misleading, for it means 'without classificatory features' and, except for the extremely nickel-rich varieties (>25 per cent Ni), they do show such features (though these are only microscopically evident). They mostly show octahedral arrangement of the kamacite but it forms discrete laths fringed by taenite, and not intersecting lamellae, as in the case of the octahedrites. These kamacite laths are set in a fine base of plessite, which forms a very significant proportion

106

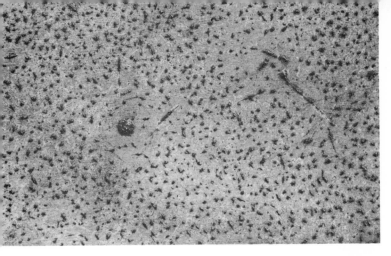

Plate 24 Cut, polished, and etched face of the nickel-rich ataxite, Warburton Range, from Western Australia

of the entire mass. There are transitional forms to the finest octahedrites, forms which display patches of intersecting lamellae and patches of ataxitic texture: again we can be certain that ataxites and octahedrites stem from the same parent body. Only very recently has a nickel-rich ataxite been observed to fall, but they may form immense masses, the largest meteorite mass known, Hoba, South West Africa, being of this class, and they are thus certainly not simply representative of very small patches within other meteoritic material.

Plate 25 Microscopic view, polished section of the Warburton Range, Western Australia, nickel-rich ataxite, showing kamacite, rimmed by taenite and set in the para-eutectoid plessite. In this iron, which contains more than 18 per cent nickel in the alloy, there is no lamellar structure; some irons, intermediate between finest octahedrite and ataxite, show rare intersecting Widmanstätten lamellae, but mainly this sort of texture. The true ataxite shows no intersecting lamellae

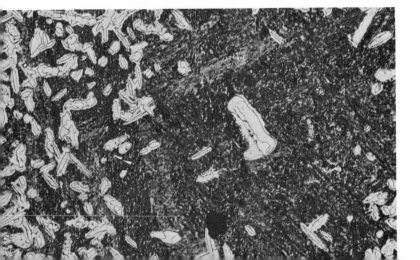

Most have less than 20 per cent nickel content, only seven exceeding this figure, with values of 23, 23, 25, 29, 33, and 62 per cent. Those with more than 27 per cent nickel content are entirely composed of taenite, the gamma nickel-iron phase.

Uhlig has suggested that this group reflects more rapid gamma–alpha phase transformation than in the case of the normal octahedrites. The whole range of iron meteorites seems to be explicable in terms of crystal fractionation in a primary melt of 11 per cent nickel content and traces of chromium, carbon, cobalt, sulphur, and phosphorus under non-equilibrium conditions, the hexahedrite phase being withdrawn first (having a composition of c 5·5 per cent nickel) to give the hexahedrite alloy: progressive nickel enrichment of the residue would ensue and the whole range of known octahedrite and ataxite alloys would come out progressively. The theoretical range of values for such a system of the early formed solid phase removed and unable to enter into subsequent reaction would be 5·5 to 6·0 per cent nickel, very close to observed values. Lovering later suggested that the 11 per cent parental melt represents the core of the parental planetary body.

The sub-solidus reaction in which the body-centred alpha phase is derived from the face-centred gamma phase occurs between 918°C and 300°C at 1 atmosphere pressure: as nickel enrichment occurs, the transformation temperature is lowered, and the two phases co-exist in equilibrium, but an alloy containing less than 6·5 per cent nickel will not retain any gamma phase below 400°C, hence the monophase character of the hexahedrites.

Widmanstätten patterns could not form except under conditions of very slow cooling in the two-phase field—kamacite forming on the octahedral planes of the original gamma phase (taenite), mimetically. Many octahedrites show a single continuous geometric structure throughout the entire meteorite. One immense crystal of taenite has been mimetically replaced by kamacite. There are, however, exceptions to this. Microscopic Widmanstätten patterns visible only with the aid of high magnification are all that has been experimentally achieved in the laboratory: the inability of experimenters to reproduce natural-scale Widmanstätten patterns in the laboratory reflects the limitation on experiment imposed by the one factor, so important in the Earth Sciences and Cosmic Sciences, and one that can never be experimentally reproduced, the effect of immense spans of time. The lack of Widmanstätten patterns in the ataxites, with more than 14 per cent nickel content, does suggest that high pressure may have lowered the alpha–gamma phase transformation temperature and diminished the extent of the alpha–gamma field in the phase diagram. Uhlig suggests that equilibrium was reached at about 300°C, no significant change being possible below this temperature because of the restriction on diffusion: he suggests a pressure of 10^5 atmospheres in the interior of the parent body. However, the question of pressure is controversial and on this point Uhlig's ideas have been disputed. There seems to be considerable evidence suggesting that high pressures are *not* required to form Widmanstätten patterns, only slow cooling; and

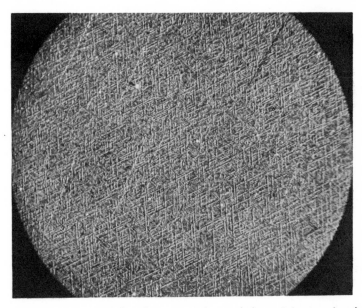

Plate 26 The strange 'ruled' etch pattern in the metal phase of the stony-iron (or iron-rich enstatite achondrite), Mt Egerton, from Western Australia. This is probably the true 'pseudo-octahedrite' pattern: it is due to the presence of an iron-nickel-silicide, 'perryite'

that meteorites were not in fact crystallised in a high-pressure environment.

The most important points that come out of the metallurgical study of meteoritic iron are the fact that the alloy is entirely compatible with a primary molten source, and that we need only a single parental source melt—possibly a single planetary core—to derive all the varied material of the iron meteorites. However, there are objections to derivation of the nickel-iron component in the stones from this single parent melt, raised by the present author in 1968 in discussion of the Bencubbin Stony Iron.

A minor variety of the meteoritic iron alloys is the type already mentioned, found in the Horse Creek Iron, and also in the metal fraction of the Mount Egerton meteorite, Western Australia, essentially a stony-iron that is a metal-rich enstatite achondrite. This metal

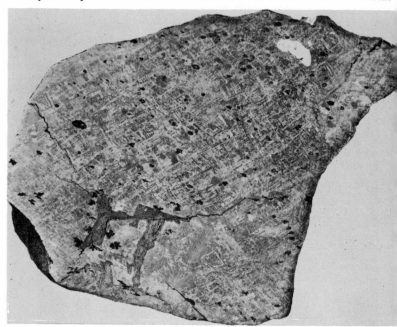

Plate 27 Brecciated phosphide-rich meteorite, Redfields, Western Australia. This iron contains 7·1 per cent nickel in the alloy but no taenite is present, only kamacite, forming a regular mesh of grains and irregular grains near the phosphide concentrations (schreibersite, in brecciation cracks). The high phosphide content has apparently inhibited Widmanstätten pattern development. The iron is mottled with dark graphite areas. The meteorite is virtually not classifiable in the accepted systems

has the composition of a hexahedrite, but displays a very fine network of fine lines on etching. It was at first thought that this material, called 'pseudo-octahedrite', was due to schreibersite inclusions, but it is now known that the structure is due to the incorporation of silicon in the metal phase, a feature of the highly reduced meteorites characterised by pure magnesian silicate minerals and low nickel metal alloy: a new mineral, perryite $(Fe,Ni)_2Si$, has been shown by E. P. Henderson to be involved (the discovery of this alloy is significant, as many authorities would like an Earth model with silicon alloyed in the nickel iron of the core). The 'pseudo-octahedrites' have nothing to do with octahedrites.

Other iron meteorites, such as the phosphide-rich iron Redfields from Western Australia, are anomalous in that their alloy 'pattern' does not fit into any classification.

Iron meteorites may show 'brecciation' textures. Such brecciation seems mainly to have occurred at an early stage, within the parent body: though there are brecciation effects due to shock on atmospheric entry and impact, these tend to be very localised, and easy to recognise: they produce a friable mass, while the early brecciation tends to be developed in a coherent, quite strongly bonded mass. Shock on collision of meteorites in Space after fragmentation of the parent body is supposed also to modify meteorite textures, in producing brecciation, but how much of this effect we can see in meteorites is a matter of conjecture. It may well be a less significant effect producing observed brecciation textures than some authorities suppose.

Other meteorites, known as 'brecciated' octahedrites, such as Four Corners (New Mexico) and Mundrabilla, Western Australia, possess a texture comprising separately crystallographically oriented octahedrite 'fragments', bonded by troilite, carbon, and schreibersite (in rare cases with silicates, olivine, and enstatite). Such iron meteorites are of extreme rarity, only eight being known. These meteorites are more correctly referred to as *granular* rather than *brecciated*. Brecciation textures are fully discussed in Chapter 13.

10: The Stony-iron Meteorites

A relatively small group numerically—less than seventy are known—these are meteorites with about equal silicate and metallic fractions. There are four main sub-divisions in the generally accepted classification, yet this classification is unsatisfactory—two of the four sub-divisions are represented by single, unique meteorites, and at least three of the meteorites commonly included in the second largest group, the *mesosiderites*, should logically be separated, on the same grounds that the unique Lodran and Steinheim meteorites have been separated to form the *Lodranite* and *Siderophyre* classificatory divisions. The author regards these three, Bencubbin, Weatherford, and Patwar meteorites, as representatives of a fifth sub-division, the *Enstatite–Olivine Stony-Irons*. Some meteorites, such as Woodbine and Netschaevo, commonly regarded as irons with silicate inclusions, also contain approximately equal silicate and metal fractions and could also be better included with the stony-irons, many of which—for example, most *pallasites*—are also, in actual fact, irons with silicate inclusions, though others, such as the Pinnaroo mesosiderite (South Australia), are silicate masses (stony) with iron inclusions.

The majority of the stony-irons are pallasites (40), the bulk of the remainder are mesosiderites (19), there are three enstatite–olivine stony-irons, one siderophyre, one lodranite, and one meteorite, Mount Egerton, which seems to be an achondrite with abnormal metal content, giving it a metal–silicate ratio typical of a stony-iron. This points to the fact that the stony-irons are, in many cases, composite meteorites composed of associations of material of more than one known meteorite type.

The Pallasites

Olivine enclosed in a nickel-iron host reticulation is the most typical textural form, but there are exceptions in which this is not the case.

The mineralogy is, however, limited by these components. There are two distinct groups according to Mason:

1 NiFe 55%   (Ni 10%)   Olivine Fa_{13} (range Fa_{10-16})
2 NiFe 30–35% (Ni 15%)   Olivine Fa_{19} (range Fa_{16-21})

In some pallasites such as Brenham, large areas of nickel-iron octahedrite composition, revealing Widmanstätten patterns on etching, are evident. This observation is compatible with the nickel-iron

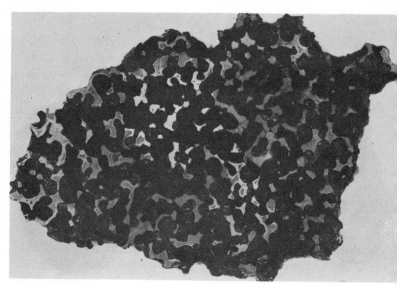

Plate 28 The Springwater, Canada, pallasite. In this meteorite the metal is set in a base of olivine (dark), but the metal still moulds the silicate grains; the texture is the normal pallasite texture, but the metal fraction is unusually sparse

ratio of the pallasites—almost all pallasites have iron of octahedrite composition forming the host reticulation. This fact, incidentally, establishes that pallasites and octahedrites stem from a common parent body, and that the host nickel-iron could stem from the 11 per cent melt that has been supposed to be parental to the whole range of iron meteorites.

Troilite, schreibersite, and farringtonite $(Mg_3PO_4)_2$ are known accessory minerals in pallasites. Pallasites may form very large masses. The Huckitta meteorite, from Central Australia, is a single mass weighing more than a ton, and the Bitberg, Germany, pallasite mass is slightly larger.

Plate 29 *The Huckitta pallasite, weighing more than 1 ton; cut, polished face. Dark olivine is enclosed in reflectant nickel-iron*

Siderophyre

There is only one example: the Steinbach, Germany, meteorite, the various pieces of which have been given separate names. Known since 1724, it is similar to a pallasite except that the silicate is an orthopyroxene, bronzite (Fs_{20}) (the orthopyroxenes form an iso-morphous solid solution series, with a complete range of composition between the two end members, the pure magnesian enstatite ($MgSiO_3$) and the iron-rich orthoferrosilite ($FeSiO_3$), which is, however, a hypothetical mineral because it is unstable in nature. The symbol 'Of' was formerly used to denote the iron silicate percentage, but of late years the symbol 'Fs' has been preferred by petrologists). Olivine

Plate 30 *The Brenham, Kansas, pallasite: octahedrite structure is indicated by the medium-coarse Widmanstätten pattern on the etched, polished, and cut surface of the metallic host to the dark olivine insets*

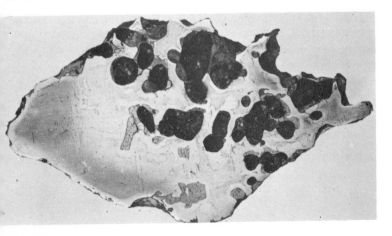

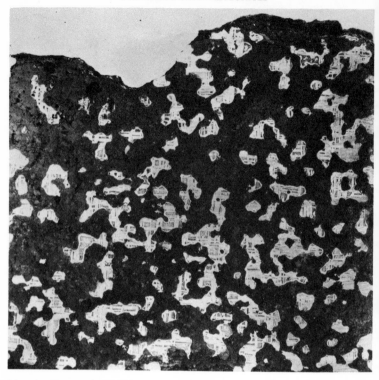

Plate 31 The Steinbach siderophyre: metal, showing an etch pattern, forms enclaves in a pyroxene base

is reported to be present in Steinbach by Tschermak in 1885. Schreibersite and chromite are accessories. The texture is coarse and closely resembles that of a pallasite.

Lodranite

Again a unique meteorite, which fell at Lodran (now in Pakistan) on 1 October 1868, is the only example: it consists of a coarse, friable aggregate of olivine and orthopyroxene grains, associated with a discontinuous aggregate of nickel-iron, which is subordinate to the silicate. The metal contains 9·4 per cent nickel. The olivine is Fa_{13} and the orthopyroxene Fs_{17}. Chromite and calcic plagioclase feldspar are reported to be sparse accessories.

Mesosiderites

Less abundant than pallasites, these stony-irons are more frequently

Plate 32 Estherville mesosiderite, showing the metal enclaves within the silicate base

observed to fall (7/2 ratio of falls to finds): there are, therefore, possibly more mesosiderites than pallasites orbiting in Space. The latter either survive terrestrial weathering better or are easier to recognise for what they are because of their striking coarse texture. The *grahamites* (carrying bytownite plagioclase feldspar) were at one time separated from the mesosiderites, which were supposed to be free from the mineral plagioclase. This term has lapsed because it has become apparent that all true mesosiderites carry appreciable calcic plagioclase in the form of bytownite or labradorite—plagioclase forms an isomorphous (solid solution) series from $NaAlSi_3O_8$, *albite* to $CaAl_2Si_2O_8$, *anorthite*. *Bytownite* (70–90 per cent anorthite) and *labradorite* (50–70 per cent anorthite) are intermediate members of this series. The nickel-iron may or may not enclose the silicates in a reticulation: though the enclosure of the silicate fraction and the converse have both been suggested as diagnostic classification

117

features of mesosiderites, it seems clear that *both* textural relation-ships are represented among the meteorites of this group.

The mesosiderites are, perhaps, the most complex of all meteorites (with the exception of the Bencubbin stony-iron), yet they must not be regarded as randomly accreted breccias cemented by nickel-iron, for each mesosiderite has a distinct mineralogical and chemical char-acter: the species of olivine, pyroxene, and plagioclase feldspar, together with the nickel-iron ratio, appear to have fixed values through-out the meteorite, and this in spite of the fact that the silicate fraction is manifestly composed both of broken crystal fragments and frag-ments of achondritic meteorite material, well known to meteoriticists in the form of discrete achondrite meteorites (the types known as *howardite* and *eucrite*).

Classifications advanced seem to have erred in treating *mesosiderite* as a 'sack-term', and not, like other meteorite classificatory terms, as

Plate 33 The Mt Padbury mesosiderite, showing the fine texture and 'agglomeratic' character; achondrite enclaves and metal enclaves or nodules, showing white, reflectant, and displaying etch patterns, are evident

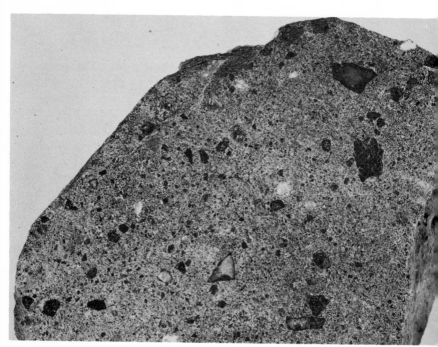

a term denoting a restricted range of mineralogical species: the minerals represented include hypersthene or bronzite (both are magnesium-rich members of the enstatite-orthoferrosilite isomorphous series, with subordinate iron content): igneous petrologists commonly define bronzite (following Poldervaart) as between Fs_{14} and Fs_{30}, and hypersthene above Fs_{30}, and it is apparent from Table 8 below that these meteorites may contain either hypersthene or bronzite in the sense of Poldervaart (though in the usage of meteoriticists it is all hypersthene). The olivine ranges from Fa_7 to Fa_{19}. The plagioclase is almost always between An_{70} and An_{90}, and is commonly around An_{85} (being within the bytownite range of the isomorphous series $NaAlSi_3O_8$ (albite)—$CaAl_2Si_2O_8$ (anorthite)). The clinopyroxene pigeonite $(Ca, Mg, Fe)_2 Si_2O_6$ may also be present (it is dominant in the Mt Padbury meteorite) and also the high temperature polymorph of silica, tridymite (SiO_2). The metal fraction is nickel-iron, ranging from $7·1$ to 10 per cent nickel content. The systematic nature of the mineralogy is shown in Table 8.

TABLE 8

Mineralogy of Mesosiderites

	Olivine	Orthopyroxene	Nickel % in metal
Crab Orchard	Fa_9	Fs_{33}	7·1
Dalgaranga	Fa_{13}	Fs_{34}	8·6
Estherville	Fa_{14}	Fs_{33}	7·1
Hainholz	Fa_{12}	Fs_{33}	8·8
Lowicz	Fa_{19}	Fs_{23}	—
Mincy	Fa_{14}	Fs_{29}	10·0
Morristown	—	Fs_{33}	7·5
Mt Padbury	Fa_9	Fs_{25}	—
Pinnaroo	—	—	8·9
Simondium	—	Fs_{33}	—
Udei Station	Fa_9	Fs_{33}	9·7
Vaca Muerta	Fa_9	Fs_{33}	7·5
Veramin	—	—	7·5

The metal fraction has the composition range of an octahedrite, and indeed at least three mesosiderites, Estherville, Mt Padbury, and Dalgaranga, contain nodules which, on etching, display octahedrite patterns. Troilite is a constant accessory in these meteorites, but is not usually aggregated with the metal phase, but rather tends to finely interlace the silicate material. The Dalgaranga meteorite has revealed howardite material, and the Mt Padbury meteorite enclaves of eucrite, diogenite, and olivine achondrite.

A recent study by Powell at Lamont Observatory, using an electron probe, suggests that there are composition ranges for silicate phases in each meteorite rather than single values, but the mineral chemistry given by Powell still suggests systematic relationships between metal and silicate phases in mesosiderites, and discounts their fortuitous aggregation together in a breccia.

The eucrite material is in every way similar to that of the extensively studied Moore County, North Carolina, eucrite. Such inclusions, with their consistent mineralogy and chemistry, cannot be dismissed as chance accumulations due to some prior process of accretion: they are part and parcel of the meteorite and must have crystallised deep within the same parent body, prior to the brecciation and metal-sulphide invasion. The ophitic texture of the eucrite, so like that of a dolerite (from terrestrial dykes), can only be explained by crystallisation from the melt, and there is also reason to believe that the diogenite has the same origin. Crystallisation in the solid state appears to have no part in such material. The three successive phases have been considered analogous to the three immiscible geo-chemical phases of Goldschmidt—*lithophile*, *siderophile*, and *chalcophile*. Though Prior attributed all mesosiderites to invasion of achondrite material by a pallasite melt (nickel-iron, olivine, and sulphide), this appears to be incorrect. The olivine seems to be the first mineral to be formed in the mesosiderites and is an early crystallisation in the silicate fraction, which was later invaded by metallic and sulphide melts after brecciation, and not by a pallasite melt. The nickel-iron in these meteorites is not, in fact, of the same chemistry as that of pallasites, the nickel percentage being consistently lower than in any of the forty known pallasites. The existence of Widmanstätten patterns in the 'ganglion-like' metallic nodules can only mean crystallisation at depth in the parent body, under conditions of very slow cooling: the achondrite material cannot surely have been brought down to such depths from higher levels in the parent body, and so we must assume that achondritic material has a similar deep provenance.

The significance of the compositional disparity between the fayalite index and the ferrosilite index, contrasting with the very

120

close correspondence in the case of chondritic meteorites, is not known.

The mesosiderites, of all meteorite types except the Bencubbin meteorite, most clearly demonstrate the unity of provenance of nearly all meteorite types. The association of octahedrite iron nodules and achondrite enclaves testifies to the fact that irons, achondrites, and mesosiderites must, at least in some cases, come from the same immediate parent planetary (or asteroidal) body.

Of late a strange hypothesis of derivation of mesosiderites by spalling of a silicate material by the agency of iron meteorites from the lunar surface has been given surprising credence in America. Among the many weighty objections to this hypothesis is the fact that both the iron and silicate fractions of mesosiderites have restricted chemical composition ranges—this appears to be quite incompatible with such a hypothesis, and the nature of crystalline material in mesosiderites seems to be quite unlike the type of congelation product likely under vacuum physical conditions at the lunar surface. The Apollo 11 recovery seems to strengthen this conclusion. Some authorities would explain these breccias by invoking shock admixture of two or more *meteorite* types—without dwelling on the objections to this idea, it may be remarked that it seems facile. Meteorite breccias like brecciated octahedrites and mesosiderites, and the Bencubbin stony-iron, are almost certainly due to processes occurring

Plate 34 Polished, etched section of the Bencubbin II meteorite, showing the coarse oriented metal reticulation devoid of etch pattern, enclosures of grey enstatite within a dark amorphous base, and a chondrite enclave (dark, right of centre)

within the parent body—long before the 'separate meteorite stage'.

The Enstatite-olivine Stony-irons

The Patwar (Pakistan) meteorite is composed of olivine (Fa_{21}) and enstatite (Fs_0), if the description published by Coulson is correct. More work on this meteorite is, however, needed since it does not seem to have been touched since Coulson's description thirty years ago.

The Bencubbin (Western Australia) meteorite, which was recovered from a field during ploughing, in 1930 (a second and larger mass being recovered close by in much the same way in 1959), is amongst the most interesting of all meteorites known, for it contains chondritic material as enclaves in a host that consists of clino-enstatite and subordinate olivine (achondritic) inset in a dark, cryptic, non-metallic base, the whole being enclosed in a metal reticulation, which, in the second mass, shows a strong directed fabric suggestive of crystallisation under directed pressure. The metal has the composition of a hexahedrite (6·6 per cent nickel), but shows no clearly defined etch pattern: troilite finely interlaces the silicate enclosures. The clino-enstatite is a pure magnesian variety (Fs_0), while the small amount of olivine associated with it is also highly magnesian (Fa_{0-7}).

The enclaves include an atypical olivine hypersthene chondrite, with the chemical composition of that class but with pigeonite the only recognisable pyroxene, olivine Fs_{19} (the value characteristic of olivine bronzite chondrites) and a spherical texture. There is a dark cryptic base and also some clear glass present in the chondrules. Another chondritic enclave is of enstatite chondrite material, with very crudely formed chondrules. Clino-enstatite, the only recognisable silicate mineral, has a composition of Fs_{0-5}, an almost pure magnesian variety. The chemical analysis is quite typical of this rare class of meteorite, but both chondritic enclaves show a small trace of carbon. An amphoterite enclave (olivine Fa_{33}) is also recently recognised. The chondritic enclaves are enveloped in achondritic material and clearly were formed before both that material and the metal solidified. The conclusion seems inescapable that the chondritic material must have formed at the same depth within the parent body as the metal and achondritic material, presumably at the level of the core mantle boundary if we can draw a terrestrial analogy: it is certainly difficult to envisage how it could be brought down to this

122

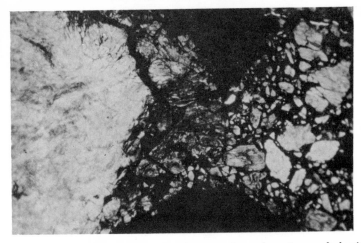

Plate 35 The Bencubbin, Western Australian stony-iron: photomicrograph showing almost pure enstatite white (left) and the same mineral inset as small fragments in dark glass (right) and dark areas of metal (top and bottom)

depth from higher levels! Also, the enclave material obeys Prior's rules in the distribution of the elements, and so does the achondritic/metallic host. This occurrence seems virtually to establish the fact that chondritic material stems from some process of crystallisation from the melt, at a very early stage in the solidification of the parent body, and that it did, in some cases, crystallise at considerable depths within it. The association is compatible with all the meteorite types represented having stemmed from crystallisation of a parent melt, of heterogeneous character, within a single parent body (though there is other very strong evidence for multiple parent bodies). The association certainly seems to dispose of Urey's theory of primary and secondary objects, and the idea that the crystallisation of the minerals in chondrules preceded the cooling of the immediate parent body. It also seems not to support the ideas of Yavnel and Mason, involving the provenance of chondrites of different types from separate asteroidal parent bodies.

Anomalous Stony-iron, Outside the Classification

The Mt Egerton (Western Australia) meteorite is unique: it was only tracked down by the author and his associates after a year's search, for though originally found during World War II, it was subsequently lost. The find of a few fragments in the collection of an

123

Plate 36 Three fragments of the unique Mt Egerton meteorite; a stony-iron or metal-rich enstatite achondrite: (a) Fragment entirely composed of metal (etched and polished); (b) Fragment composed of metal, enclosing abundant but subordinate enstatite (white); (c) Fragment composed of enstatite (single crystal, mottled white and black) together with small reflectant enclaves of metal

elderly lady (widow of one of the finders), who died a week later, led to a major recovery of fragments. Much of the material is predominantly composed of silicate, but smaller lumps are of metal enclosing silicate. Because only lumps a centimetre or so in diameter have been recovered, the proportions in the mass are not known, but they may well have been between 2/1 and 3/1 silicate to metal ratio. The metal has pseudo-octahedrite character and contains the mineral perryite ($FeNi_2Si$). The silicate consists of crystals up to several centimetres long of pure-enstatite (Fs_0, but *not* clino-enstatite), and has a 'pegmatoid' character. The meteorite could be regarded as an achondrite (enstatite achondrite or aubrite) with metallic inclusions, but much of the material recovered is metal with silicate inclusions. It is interesting to note that the chemistry of the mineral phases (Fs_0 and 6.38 per cent nickel) is entirely consistent with Prior's Rules (Chapter 13).

11: The Stony Meteorites

Chondrites

The chondrites—stony meteorites which display chondritic fabric, a spheroidal aggregation of minerals unlike any known in terrestrial rocks—account for the greater part of the known stony meteorites of which 1,152, plus 62 doubtful or paired occurrences, are listed by Hey. A very few of the carbonaceous and enstatite chondrites display no recognisable chondrules, but on account of their chemical and mineralogical affinities, are classified with the chondrites. Conversely, rare chondrules are reported from howardites and mesosiderites. The distinction between chondrites and achondrites is nowadays based more on chemical and mineralogical character, than on the original criteria of presence or absence of chondrules. The peculiar chondritic fabric is probably no more than a reflection of crystallisation under unfamiliar physical conditions—possibly involving degassing on a grand scale.

There is one group of chondrites—the amphoterites—that was for many years referred to the achondrite group. Recent investigations by Kvasha, Mason, and Wiik have, however, shown that they are olivine-hypersthene chondrites showing unusually intense recrystallisation, and containing an unusually ferroan variety of olivine.

Chondrites were subdivided into thirty-one groups by Brezina in 1904. This classification, based on superficial differences of no taxonomic significance, is now followed by few scientists. The first realistic classification was Prior's, improved on later by Mason.

Prior, basing his classification on mineralogy and chemistry, erected three classes of chondrites: Mason added the olivine-pigeonite and carbonaceous chondrites, but the first of the new classes has not found general acceptance, these additional carbonaceous types all being grouped as carbonaceous chondrites, and the olivine-pigeonite chondrites being referred to type 3.

127

The criteria adopted by Prior to distinguish his three classes were:

1 MgO/FeO molecular ratio in bulk analysis.
2 MgO/FeO molecular ratio in silicate fraction.
3 Fe/Ni ratio in metallic fraction.

In practice, his second and third criteria are used. The ranges for the second value are:

Olivine-hypersthene chondrites:	2–4	common chondrites
Olivine-bronzite chondrites:	4–9	
Enstatite chondrites:	$\leqslant 9$	

It has been found that the amount of metal present in a chondrite, varying as it does sympathetically with these ratios, gives an easy, quick indication of the type of chondrite, applicable to cut surfaces of hand specimens. Enstatite chondrites contain about 25 per cent nickel-iron, bronzite chondrites about 15 per cent, and hypersthene chondrites about 5 per cent. It must, however, be noted that a few of the rare unequilibrated chondrites contain negligible metal, whatever their group.

The lack of accurate analyses for many chondrites (and the pointlessness of analysing them due to weathering) led Mason to evolve a simple instrumental method of determining to which class a chondrite belongs, based on the use of an X-ray diffractometer which measures crystal lattice dimensions. The fayalite index of the olivine is measured, and thus this method is not applicable to enstatite chondrites, which rarely if ever contain olivine, or to unequilibrated or carbonaceous chondrites which show wide internal variation within single olivine grains. The method can be checked by optical determinations of refractive indices, but, despite the variation in texture revealed in thin sections of chondritic material under the microscope, mineral determinations by optical methods tend to be difficult due to shock modifications and technical difficulties in preparing thin sections resultant on the presence of malleable metal specks in the silicate aggregate.

Mason gives the critical indices for the four chondrite classes easily determinable by diffractometry:

	Olivine	Orthopyroxene
Enstatite chondrites	(rarely, if ever, present)	Fs_0

Bronzite chondrites	Olivine Fa_{15-22}	Orthopyroxene is of about equivalent value Fs index to the olivine
Hypersthene chondrites (includes amphoterites)	Olivine Fa_{22-32}	
Carbonaceous chondrites Type 3	Fa_{32-40}	(not present)

Enstatite Chondrites

Only fourteen are known, of which nine are falls. These rare chondrites consist of crystalline, pure or almost pure magnesian pyroxene, enstatite or clino-enstatite: 20–28 per cent metal (Fe/Ni = c 13), entirely kamacite, in composition equivalent to hexahedrite alloys: troilite (7–15 per cent): and minor accessories—diopside (a calcic clinopyroxene) exsolved from the enstatite, quartz, tridymite, or cristobalite (indicating an excess of silica in the composition, and hence the absence of the silica-undersaturated mineral olivine). Quartz, so abundant in terrestrial rocks, only occurs in this meteorite class (St Marks, South Africa; Abee, Canada). Clino-enstatite is probably a shock modification of pure enstatite, which also occurs in such meteorites. These meteorites are in a highly reduced state, the iron being restricted to the metal and sulphide phases, and calcium, manganese, and chromium all occurring in the sulphide (oldhamite, alabandite, daubreelite). Even silicon may occur in the metal phase (Ringwood). Densities vary from 3·45 to 3·66. Chondrules are poorly formed or absent in most cases, but St Marks is an exception, containing well-formed enstatite fan chondrules. Blithfield (Canada), Hvittis (Finland), and Pillistfer (Estonia) show no chondrules at all.

Olivine–bronzite Chondrites

No less than 405 of these (with five doubtful or paired fall occurrences) are listed by Hey: of these 225 were seen to fall. They consist of olivine and bronzite in about equal proportions: possibly a little clinohypersthene: and about 15 per cent metal corresponding to the composition of the alloy of octahedrites. Troilite and oligoclase are common accessories (up to 10 per cent). The density varies from 3·6 to 3·8, in unweathered specimens. They form, with the hypersthene chondrites, the prolific group known as the *common chondrites*. They grade compositionally and mineralogically into the hypersthene

129

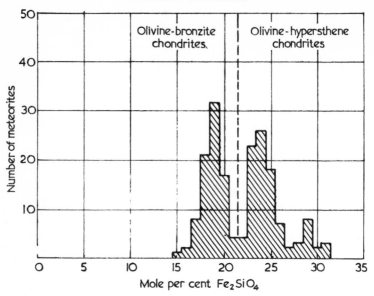

Fig 16 Olivine composition in 180 olivine-bronzite and olivine-hypersthene chondrites, determined by X-ray diffraction (from Mason, 1962)

chondrites, but two distinct population maxima define the two groups statistically. There is a chemical and mineralogical discontinuity between these chondrites and the rare enstatite chondrite class.

The plagioclase, oligoclase (10–30 per cent anorthite) is modally evident in examples that have undergone thermal recrystallisation (Mason), but the plagioclase component is held in the cryptic opaque (part glassy?) base in the spherical, unrecrystallised examples displaying perfect chondrule forms. Some of this cryptic plagioclase component seems also to be converted to the lamellar calcium-bearing silicate of the clino-pyroxene group, evident in such unrecrystallised chondrites.

Calcium, manganese, and chromium are present in oxide form, not as sulphides, and a portion of the iron is in oxide form within the ferromagnesian silicate phases. These meteorites are thus more oxidised than the enstatite chondrites. The abrupt cut-off of iron from the silicate phases below 14–15 per cent content is not understood: it contrasts with the smooth gradation between hexahedrites and octahedrites, the metal alloy classes that are compositionally equivalent to the accessory metal areas in these two chondrite classes. The cut-off must be due to some obscure internal limitation on the particular physico-chemical system involved, rather than to any cosmic factor such as provenance from different parent bodies.

Plate 37 The Cocklebiddy olivine-bronzite chondrite, a dark stone recovered in a remarkably fresh state on the Nullarbor Plain in Western Australia. Note the higher metal content than in the Lake Brown meteorite in the next photograph

The olivine–bronzite chondrites display some of the most perfect chondritic fabrics of any meteorites, and also commonly contain glass, within the chondrules and in the interstices between the chondrules. This is interesting, for these meteorites must be of great age (Chapter 18) and ancient glass is rare in very old terrestrial rocks, such glasses suffering devitrification (though Archaean glasses are now being recognised in Western Australia): the presence of clear glass in chondrites suggests that passage of immense periods of time is not enough, on its own, to effect devitrification.

Olivine–hypersthene Chondrites

With 492 listed by Hey (plus twelve doubtful or paired fall occurrences) this is the largest of all meteorite classes: 312 of these were

131

seen to fall. They consist of olivine, dominant, and orthopyroxene—commonly referred to as hypersthene, but within the commonly accepted definition of Poldervaart, actually within the bronzite

Plate 38 The Lake Brown, Western Australia, olivine-hypersthene chondrite, showing the dark fusion crust and reflectant metal specks sparingly evident within the grey silicates on the cut face

range, except for the Fs_{30-32} varieties occurring in some amphoterites. A calcic clinopyroxene showing lamellar twinning under the microscope is a common accessory. The accessory nickel-iron is an aggregate of kamacite and taenite, though separation into discrete areas of the alpha and gamma phases is not always detectable without the aid of the optical microscope. Other accessories are troilite and oligoclase (up to 10 per cent). The troilite tends to be aggregated with the metal, as in all chondrites.

These chondrites are even more highly oxidised than the bronzite chondrites, most of the iron being contained in the oxide form within the ferromagnesian silicates.

Though the calcium-bearing clinopyroxene was figured by Tschermak in 1885, the normative calcium silicate evident in the chemical analysis of such meteorites was long thought to be due to fine 'Schiller' inclusions of exsolution origin within the orthopyroxene grains. These clinopyroxene grains are so fine that they were overlooked in mineral counts, made, for obvious reasons, on the larger grains (Mason).

Densities range from 3·5 to 3·6 for unweathered specimens.

Carbonaceous Chondrites (*Type 3*)

Two separate classes were defined by Brezina—ornansites and spherulitic carbonaceous chondrites of type 3, but these seem to form one class of chondrite. They are not all markedly carbonaceous, despite a typical black coloration, the carbon content being usually less than one per cent. Only a trace of carbon seems to be required to produce such dark coloration of meteoritic stones. Clarke, Jaresowich, Mason, and Nelen list fourteen, including the Allende, Mexico fall of masses totalling more than 1,000 kg. All but two are falls. These chondrites contain abundant olivine (70 per cent), only about 5 per cent pigeonite, no orthopyroxene, plagioclase (5–10 per cent),

Plate 39 Cut, polished surface of the Vigarano, Italy, carbonaceous chondrite type 3. Well-formed chondrules are set in a dark, carbonaceous base

troilite (5 per cent) and nickel-iron only sparsely present (0–5 per cent). The nickel-iron is highly nickeliferous, being entirely composed of taenite, and thus being equivalent to the alloy of ataxites. The sulphide mineral phase may be the nickel-iron sulphide pentlandite instead of troilite. The total amount of iron present, decreasing progressively through the three classes described above, is higher than in the previous class, which is, perhaps, surprising:

Total iron (in silicate, sulphide, and metal phases):

CEn	CBr	CHy	Cc Type 3
23–35	27–30	20–23	24–26

The olivine in all but the three slightly recrystallised $Cc3_{47}$ meteorites is of variable composition within the limits of individual grains and from grain to grain. This feature, similar to the variability of the unequilibrated chondrites, indicates incomplete attainment of chemical equilibrium, possibly, in this case, due to a protective 'armour plating' of metal around chondrules. The silicates contain between 30 and 40 per cent FeO in the base oxides (Ca:Mg:Fe). It has been suggested that the pigeonite reflects unusually high calcium content, not a high temperature inversion of unstable hypersthene to pigeonite (another possibility). Mason believes that the small amount of pyroxene crystallised takes up all the calcium, the pyroxenes individually holding more calcium than in the case of the pyroxenes of the common chondrites, which are more abundant but can incorporate less. It could be that silica saturation is the real controlling factor: these meteorites contain more iron and less silica, and this fact places a restriction on the amount of the silica-saturated mineral pyroxene that could form. The plagioclase is also more calcic than usual in some cases (An_{50} in Mulga West).

Lately, a *Vigarano* (A) and *Ornans* (B) sub-type have been erected by van Schmus. The former type is characterised by spongy chondrules set in a fine opaque matrix, the chondrules containing inclusions of iron or sulphide. Type B is characterised by small, metal-poor, closely-packed chondrules. Meteorites of this sub-type may contain peculiar rimmed aggregates of gehlenite, anorthite, augite, spinel, nepheline, sodalite, grossular, and perovskite—products of glass devitrification.

The density of the carbonaceous chondrites of type 3 ranges from 3·6 to 3·8 in unweathered meteorites, and they are characterised by well-formed, spherical chondrules.

Plate 40 The Cold Bokkeveld carbonaceous chondrite type 2, showing the dark carbonaceous fusion-crust coated mass

Ranging from the enstatite chondrites to the type 3 carbonaceous chondrites, one can see a pattern emerging of immiscible silicate and metal fractions (lithophile, siderophile), showing compositional variations through a discontinuous and continuous range respectively. The oxidation state ranges from reduced to highly oxidised, with more and more iron going into the silicate fraction. The silicate assemblage of the enstatite chondrites displays silica oversaturation, whereas the remainder of the chondrite classes described display undersaturation, progressively more marked: in the devitrified glass of the type 3 carbonaceous chondrites we even see the mineral phases of the comparatively rare peralkaline terrestrial igneous rocks. The discontinuities evident in the silicate fraction may be attributed to the complex and discontinuous nature of the silicate systems that govern crystallisation of pyroxenes and olivines.

135

Carbonaceous Chondrites, Types 1 and 2

These are few in number—only eighteen are recorded. All are falls. It is doubtful whether the fissile and easily decomposed material that comprises these stones could survive on the ground for any length of time, even in an arid climate. This material is also of nondescript

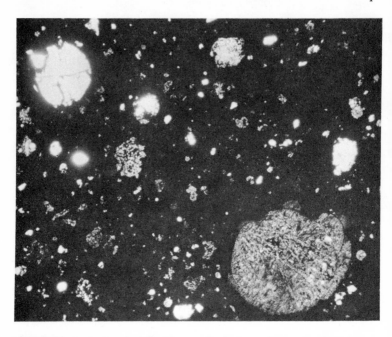

Plate 41 Photomicrograph of the Murray, Kentucky, carbonaceous chondrite type 2, showing small chondrules and grape-bunch clusters of mineral grains set in a dark, carbonaceous matrix

appearance and so not liable to attract attention. Very small amounts of material have been recovered from these falls: Murray, Kentucky (1950), 12·6 kg—the largest recovery recorded—and at the other end of the scale, Al Rais, Arabia (1957), 0·16 kg; Bells, Texas, 0·3, kg; Crescent, Oklahoma (1936), 0·08 kg; Erakot, India (1940), 0·11 kg; and Tonk, India (1911), 0·01 kg. Detailed study is limited by such small stores of material to work on, and most of the research to date has been concentrated on a handful of the larger recoveries.

These meteorites have, nonetheless, been more intensively studied by scientists than any other single group. They have been the subject of two controversies during the last two decades, concerning their possible content of primitive fossils and their possible parental status to other forms of chondrite. These subjects are treated separately in Chapters 14 and 17.

They are characterised by low density and significant amounts of hydrous minerals. The two types are defined according to Table 9.

TABLE 9

Carbonaceous Chondrites, Types 1 and 2

Type	Density	Average chemical composition					Mineralogy
		SiO_2	MgO	C	H_2O	S	
1	2·2	22·56	15·21	3·54	20·08	6·20	Amorphous hydrous silicates and sulphates: finely divided iron-nickel spinel (highly magnetic): no chondrules: little or no iron sulphide or free nickel-iron.
2	2·5 to 2·9	27·57	19·18	2·46	13·35	3·25	Sheet minerals, serpentines, and chlorites are the principal minerals: the mass is weakly magnetic or non-magnetic: free sulphur is present: chondrules of olivine (almost pure forsterite) and enstatite or clino-enstatite are present: the olivine and the pyroxene are of variable chemistry: little or no iron sulphide, or free nickel-iron.

The values in this table are due to Mason, differing slightly from those of Wood (Chapter 12). Black, thick fusion crusts are characteristic of these two types: fresh cut surfaces of the interior of the masses appear black, commonly with a greenish tinge. Type 2 carbonaceous chondrites appear like dark-coloured normal chondrites, but masses of type 1 are very difficult to recognise as meteorites at all.

Mason has noted that chemical analyses of these masses, recast to atomic percentages excluding water, carbon, oxygen, and sulphur (as suggested by Wiik), become uniform despite their variability of content of these components: however, in addition, their content of inert gases and the volatile elements bismuth, lead, thallium, and mercury, distinguishes them from other chondrite material. Despite these differences, they could be reasonably regarded as products of extreme oxidation of chondrite material.

Achondrites

The name 'achondrite' implies an absence of chondrules, but there are exceptions—howardites, as we have noted, do in rare cases manifest chondrules. Some of the achondrites display textures and

137

Plate 42 The Orgeuil, France, carbonaceous chondrite type 1, showing the dark fusion crust, and friable nature of the mass in broken surface. White areas of hydrous minerals are clearly visible

Plate 43 The Aubres, France, enstatite achondrite, showing the brecciated character of the crystalline aggregate, composed largely of white clino-enstatite crystals, which loom through a translucent light-coloured fusion crust

mineralogy astonishingly similar to those of terrestrial dolerites and basalts. The term eucrite is in fact applied to both achondrite meteorites and terrestrial igneous rocks of gabbroic character of similar mineralogy. The howardites and eucrites are loosely termed the basaltic achondrites, and they comprise, together with the unique angrite and the two known nakhlites, the calcium-rich achondrites. The calcium-poor achondrites include the enstatite achondrites (aubrites), hypersthene achondrites (diogenites), olivine–pigeonite achondrites (ureilites) and the unique olivine achondrite (chassignite). The separation of these two groups reflects a marked compositional contrast—the calcium-rich group has a calcium content ranging between 5 and 25 per cent, whereas the calcium-poor group has a range from 0 to 0·3 per cent.

Achondrites are comparatively rare—sixty-three in all, nearly all falls, are listed by Hey. This comparative rarity must be real: they cannot be abundantly represented among the meteoroids orbiting the Sun and liable to intersect the orbit of the Earth. The lack of finds of meteorites of this group indicates that, like other meteoritic stones, they weather rapidly after arrival on Earth, and are not easy to detect, unlike irons. The chassignite resembles closely the terrestrial igneous rock, dunite; the diogenites the igneous pyroxenites; and the howardites and eucrites the common igneous rocks, dolerite and basalt.

Calcium-poor Achondrites

Twenty-one of these are known.

Enstatite Achondrites

Eight falls and one find are listed by Hey: MgO and SiO_2 comprise virtually the entire silicate fraction, represented by pure enstatite, possibly intergrown with clino-enstatite. The magnesian olivine forsterite may be present as an accessory, and diopside is also recorded. Oligoclase may also be present in accessory amounts. The nickel-iron contains about 6 per cent nickel in the alloy, again being about equivalent to the alloy of hexahedrites (cf enstatite chondrites): it is entirely composed of the alpha phase, kamacite. Troilite, oldhamite, and schreibersite are recorded as accessories: the presence of oldhamite indicates a similar highly reduced state to the enstatite chondrites. Similarly, Pesyanoe, USSR, contains a magnesium sulphide. Of the nine recorded occurrences of meteorites of this class, only one, Shallowater, Texas, is not

139

Plate 44 The Dingo Pup Donga, Western Australia, ureilite showing the fine-textured, dark, carbonaceous cut surface of the sole mass, and weathered fusion crust

brecciated: it is composed of enstatite crystals up to 45 mm long. The meteorite, Mt Egerton, described in Chapter 10, appears to be a similar aggregate of large enstatite crystals, but with an unusually high metal content for an enstatite achondrite. The coarseness of these crystal aggregates makes them subject to brecciation on the application of any stress, the mass being friable. The term whitleyite was proposed for the Cumberland Falls, Kentucky, meteorite of this class on account of its unusual polymict breccia character, but this term is superfluous.

Olivine–pigeonite Achondrites

Of these, only five are known, three being finds and two falls. They show some resemblance to the carbonaceous chondrites of type 3. The nickel-iron content may be greater than in other achondrites, for the group as a whole is characterised by paucity of metal. The nickel content of this metal is extremely low in the first three ureilites

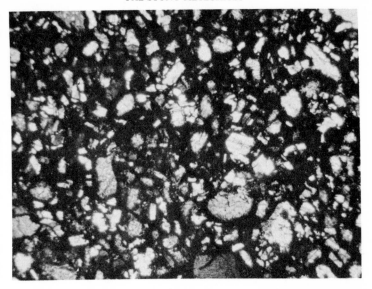

Plate 45 Photomicrograph of the Dingo Pup Donga ureilite showing olivine and lamellar pigeonite grains set in a dark carbonaceous matrix

recovered—Dyalpur and Goalpara, India, and Novo Urei, USSR—all three of which contain appreciable metal. In the latest two ureilites recovered—North Haig and Dingo Pup Donga, Western Australia—there is only a very slight amount of metal, and it has a quite high nickel content. Mineralogically, the ureilites are all very similar, consisting essentially of olivine and pigeonite set in a dark, carbonaceous base. In some, the nickel-iron takes the form of granules fringing the silicates. Troilite is an accessory. Graphite and diamonds have been recognised in four of these meteorites. In Novo Urei the olivine is recorded as Fa_{21} and the pigeonite as $Fe_{20}(Mg,Ca)_{80}$; in North Haig, the olivine is variable, Fa_{0-23}: in Dingo Pup Donga it is Fa_{10}. The resemblance to the carbonaceous chondrites of type 3 is mainly in the chemistry, but there is also a mineralogical similarity in the mineral phases present, the variable olivine, and the presence of carbon. The high nickel content of the metal of North Haig and Dingo Pup Donga is another point of similarity. The olivine is, however, much less ferroan and the pigeonite present in much greater abundance.

141

Olivine Achondrite

This unique meteorite fell at Chassigny, France, in 1815. A 4 kg stone, it shows the structure of a terrestrial dunite, a monomineralic olivine rock of igneous origin. The olivine makes up 90 per cent of the mass, and is Fa_{33}, a quite ferroan variety. There is a trace of plagioclase (oligoclase?) and nickel-iron, very rich in nickel. It has been suggested that this meteorite is a product of thermal recrystallisation of the carbonaceous chondrites of type 3, its chemistry being not dissimilar (Mason). Yet the calcium-poor achondrites, in spite of their similar chemistry and mineralogy, do not seem to be simple thermal recrystallisation derivatives of chondrites, their textures being very different from those of the most recrystallised chondrites, and certain chemical differences suggest that they do not stem from an intermediate chondrite stage of crystallisation but directly from a melt not entirely dissimilar to the parental melt from which the chondrites crystallised.

Hypersthene Achondrites

Eight are known, all falls. The principal mineral component is

Plate 46 Cut and polished surface of the Johnstown, Colorado, hypersthene achondrite showing the crystal breccia texture of crystalline enclaves set in a comminuted matrix

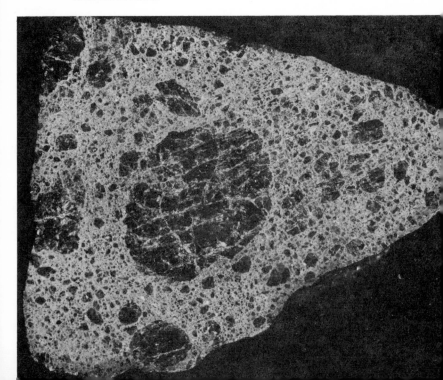

Plate 47 Photomicrograph of the Johnstown, Colorado, hypersthene achondrite showing the crystal breccia microtexture

orthopyroxene (Fs_{25-27}), strictly bronzite according to the now widely accepted Poldervaart classification of pyroxenes. These meteorites are almost monomineralic. Chromium is a minor elemental component of the pyroxene of Johnstown, Colorado, and Shalka, India. The same feature is seen in the pyroxene of the enclave of this achondrite material in the Mt Padbury, Western Australia, mesosiderite. Oligoclase, olivine (Fa_{27-28}), chromite, and troilite are accessories. A small amount of nickel-iron may be present. Tridymite has been identified in Ibbenbuhren, Germany, and in the Mt Padbury enclave. These achondrites, like the howardites, eucrites, and mesosiderites, tend to have compositions close to the line of silica saturation, and a slight shift either way introduces olivine or tridymite. Only the meteorite Tatahouine, Tunisia, is unbrecciated: again, the coarseness of the primary crystal grain aggregates renders these meteorites friable and liable to brecciation. The crystal grains tend to show intense optical strain. Chemically these meteorites are uniform, but there is a marked gap between their

143

composition and that of the enstatite achondrites—a gap even more marked than that between the composition of the bronzite and enstatite chondrites.

The Calcium-rich Achondrites

The pyroxene–plagioclase achondrites make up the greater part of this group. The essential compositions are:

Eucrites: pigeonite + bytownite–anorthite:
              FeO/MgO + FeO, 48–66 per cent
Howardites: hypersthene + bytownite–anorthite:
              FeO/MgO + FeO, 29–40 per cent

Plate 48 The Juvinas, France, eucrite, showing the glossy black fusion crust, and grey achondritic broken surfaces

Twenty-six eucrites (twenty-three falls) and thirteen howardites (twelve falls) are recorded by Hey, and two new eucrite recoveries have been recently made in Australia (Millbillillie, Western Australia, and another from Victoria, both falls).

The suggestion of Lacroix that howardites are no more than brecciated eucrites is incorrect, for there are clearly two distinct chemical and mineralogical groupings. The chemistry is quite uniform in each group, the only unusual analysis being obtained from Sierra de Mage, Brazil, the high CaO, Al_2O_3 and low MgO, FeO values being simply due to the fact that it approaches monomineralic anorthositic composition, plagioclase being dominant in the mode (Mason). The mineralogy is:

Eucrite: pigeonite $\pm$ hypersthene, calcic-plagioclase, olivine or tridymite: ilmenite, chromite, magnetite, nickel-iron, and troilite.

Howardite: hypersthene $\pm$ pigeonite, calcic plagioclase, olivine or tridymite, ilmenite, chromite magnetite, nickel-iron, and troilite.

The presence of very slightly over- and under-saturated examples with respect to silica may be compared with the oversaturated tholeiitic and undersaturated alkalic basalts that together comprise the bulk of the Earth's lavas.

Textures are characteristically ophitic or sub-ophitic, like those of terrestrial dolerites. There is a record of a preferred mineral orientation (*petrofabric*), in Moore County, North Carolina according to Henderson, and both the Mt Padbury eucrite enclaves (in mesosiderite material) and the new Millbillillie eucrite show layered structures, in the latter case resembling magmatic layering. Such structures could indicate that the asteroidal parent body possessed a gravitational field similar to that of the Earth, at the surface. Yet it is difficult to equate this evidence with an asteroid-sized parent body, and the fact that other evidence suggests that achondrite material comes from deep within the parent body.

Both howardites and eucrites characteristically show brecciation: some eucrites are not brecciated, however. Shergotty, India, has given rise to the superfluous name, Shergottite: it is a eucrite intensely shocked and containing feldspar laths converted to maskelynite, and thus anomalously isotropic under the microscope. The feldspar was, in this case, labradorite, an unusual variety in meteorites, though ubiquitous in terrestrial igneous rocks.

Augite Achondrite

The sole angrite fell at Angra dos Reis, Brazil, in 1869: unfortunately, much of the original recovery has been lost, but the surviving fragments show it to consist largely of augite (90 per cent), together with some olivine and troilite. This is the only meteorite to contain primary augite, the only other occurrence being in the form of minute grains within secondary devitrification products. This mineral is, in contrast, the most common pyroxene in terrestrial igneous rocks. There are many early records of augite in meteorites, but these refer to pigeonite, not augite. The calcium oxide content of the angrite is 24·51 per cent and the titanium dioxide content 2·39 per cent—values higher than in any other meteorite. The chemical composition of this meteorite is compatible with its uniqueness, for it marks the extreme in stony meteorite composition.

Plate 49 The Nakhla, Egypt, diopside-olivine achondrite showing the black, shiny fusion crust on one of the masses

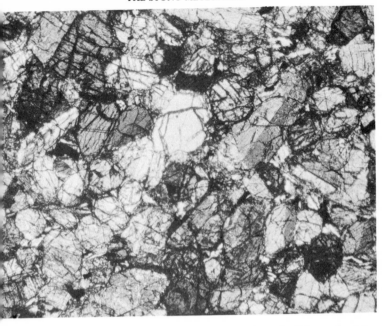

Plate 50 Photomicrograph of the Nakhla, Egypt, diopside-olivine achondrite, showing the achondritic aggregate of twinned diopside and untwinned olivine grains, with minor amounts of lamellar twinned plagioclase in the interstices

Diopside–olivine Achondrites

Two of these are known, one a fall (Nakhla, Egypt, 1911) and the other a find (La Fayette, Indiana). The former was a fall of about 40 stones totalling 40 kg, whereas the latter weighs only 0·6 kg. These meteorites approach Angra dos Reis in their extreme calcium oxide contents (15·17 per cent in Nakhla), and again their rarity is compatible with their extreme chemistry. The minerals present are the calcium-rich clinopyroxene diopside (75 per cent), olivine (15 per cent), plagioclase (oligoclase–andesine) in accessory amounts, and magnetite. The texture is locally ophitic, like that of the eucrites and howardites, resembling that of terrestrial dolerites. The specific gravity is 3·47. The olivine of the Nakhla stone is Fa_{66}, by far and away the most ferroan found in any meteorite, and again indicative of the extreme chemistry of these achondrites.

147

12: The Physical Properties and Chemistry of Meteorites

Physical Properties (based largely on Wood, 1963)

Mechanical

Specific gravity: The specific gravities of various meteorite classes are given in Table 10.

TABLE 10

Specific Gravities of Meteorites

Class		Sp gr
Stones (chondrites)	Carbonaceous 1	2·2
	Carbonaceous 2	2·5–2·9
	Carbonaceous 3	3·4–3·6
	Enstatite	3·45–3·66
	Bronzite	3·39–3·90 (av 3·66)
	Hypersthene	3·29–3·62
Stones (achondrites)	Eucrites	2·95–3·36 (av 3·26)
	Howardites	(av 3·23)
	Aubrites	3·35*
	Diogenites	3·20–3·41
	Ureilites	3·05–3·27
	Nakhlites	3·47
	Angrite	(3·25–3·50)*
	Chassignite	3·25*
Stony-irons	Pallasites	(av 4·74, 4·90)†
	Siderophyre	4·90
	Lodranite	4·60
	Mesosiderites	4·85 (max 6·2)
	Enstatite–olivine (Bencubbin)	5·32
Irons	Hexahedrites	7·77
	Octahedrites ⎰ Og, Ogg	7·81
	⎰ Om	7·81
	⎱ Of, Off	7·77
	Ataxites	7·85

* Based on values for mineral phases.
† Two averages obtained by different authorities.

Porosity. Some chondrites display spongy texture, their percentage of pore space being quite high. Values listed by Wood range from 18·20 per cent down to 7·08 per cent for spongy chondrites. An enstatite chondrite, Pesyanoe, USSR, has 15·10 per cent pore space. This characteristic of stony meteorites suggests that they cannot be products of crystallisation in high-pressure environments, deep within parent bodies of planetary size.

Crushing strength. Compressive strengths vary from 3,810 kg/cm² (hypersthene chondrite, La Lande, New Mexico): through 1,627 kg/cm² (bronzite chondrite, Morland, Kansas) and 844 kg/cm² (bronzite chondrite, Ness County, Kansas) to 63 kg/cm² (hypersthene chondrite, Holbrook, Arizona); and the Bjurbole, Finland, spherical hypersthene chondrite is even weaker than the Holbrook mass, being easily crushed between thumb and forefinger.

An octahedrite iron, Descubridora, Mexico (Om), has a crushing strength of 3,800 kg/cm², equalling that of the strongest of the chondrites measured: the Khor-Temiki, Sudan, enstatite achondrite is even more friable than the Bjurbole chondrite, and fragments tend to fall apart, disintegrating into a loose aggregate of crystals on the application of the slightest pressure: they are commonly exhibited in glass bottles for this reason.

Elastic wave velocities. Russian scientists have studied this property, obtaining values for eight chondrites from 2,050 to 4,200 m/s for longitudinal waves (average 3,420). For transverse waves they range from 600 to 1,220 m/s. The elasticity of meteoritic stones is, as a rule, appreciably lower than that of terrestrial crystalline rocks.

Thermal

Heat capacity. This property is largely dependent on metal and troilite content in a particular meteorite. Again this property has been studied by Russian authorities, who obtained values for six chondrites ranging from 0·166 for the bronzite chondrite, Misshof, Latvia, of 23 per cent by weight metal and troilite content to 0·240 for the hypersthene chondrite, Kukschin, USSR, which contains only 8 per cent by weight of metal and troilite.

Thermal conductivity. This property is similarly dependent; four stones determined by Russian workers gave values ranging from $3·6 \times 10^{-3}$ Cal/Sec Cm Degree C for the enstatite achondrite, Norton County, Kansas, with 2 per cent by weight of metal and troilite, to $5·8 \times 10^{-3}$ for the bronzite chondrite, Svonkov, Ukraine.

Thermal conductivities of meteorites are, as a rule, low—hence the behaviour of stones, stony-irons and irons in developing only thin melt skin zones due to extreme heating of frictional origin on entering the Earth's atmosphere.

Melting points. Melting points of the principal mineral phases present in meteorites are:

Olivine—	pure forsterite (Mg_2SiO_4)	....1,890 °C (±20)
	hyalosiderite (Mg, Fe)$_2$SiO$_4$	....1,430 °C–1,215 °C
	pure fayalite (Fe_2SiO_4)	....1,100 °C
Orthopyroxene—	pure enstatite ($MgSiO_3$)	....1,554 °C
	bronzite (Mg, Fe)SiO$_3$	....1,459 °C–1,360 °C
	pure ferrosilite ($FeSiO_3$)	....1,100 °C
Native iron—	artificial, pure (Fe)	....1,533 °C
Troilite (FeS)		.... 743 °C

Viscosity. Iron meteorites are plastic or viscous, and malleable in the cold state, though they cannot be tempered without admixture of terrestrial iron. Ancient peoples did utilise meteoric iron to make weapons, for the memoirs of Jehangir relate that the fall of a two-kilogramme meteorite in India caused the author to order a smith to forge a knife and sword from it, but the smith had to mix one part of terrestrial iron with three of meteoric iron to achieve this. Stony meteorites have a greater plasticity or viscosity than terrestrial ultra-basic and basic rocks which they resemble in composition, because of their metal content.

Electrical Properties

Resistivity. The resistivity of stony meteorites is two or three orders lower than that of equivalent terrestrial ultrabasic and basic rocks, and this, again, is likely to be due to their metal content. Variations in metal content correspond to variations in resistivity values. Values for chondrites range from 6.9×10^{-9} ohm cm to 5×10^{-8} ohm cm.

Magnetism. Residual magnetism has been reported in meteorites in the case of a few irons and stones. Its origin is obscure, but it might indicate that the parent body possessed its own magnetic field. Magnetic susceptibilities vary with metal content: for chondrites values ranging from 0.87 to 2.56×10^{-2} cgs units have been recorded.

Optical Properties

Albedo (*reflectance*). Chondrites reflect evenly over the entire spectrum, being essentially grey reflecting bodies, with a slight degree of redness attributable to terrestrial oxidation of unstable minerals that yield iron oxides. Black chondrites only reflect 4·5 per cent of the incident light, and have thus a much lower albedo than the grey chondrites. In contrast, enstatite achondrites, characteristically very light coloured, reflect 44–50 per cent of the incident light. Light rather than dark grey predominates in the wide range between these two extremes. Krinov compares colour indices of meteorites and members of the Solar System for which values are based on comparisons of visual and photographic stellar magnitudes:

Sun	0·79
Moon	1·08
Meteorites	1·08
Asteroids	—wider scatter of colours, producing bluer colours than meteorite masses*
Terrestrial rocks	—brown colours predominate, due to oxidation effects.

Chemistry of Meteorites

Sampling of meteorites presents problems, for they are mostly inhomogeneous and pure mineral phases are difficult to separate from them. Also in unequilibrated meteorites the mineral phases may show variations within the limits of a single thin section. Certain components such as copper, which occurs in chondrites as native copper, may be so sparsely disseminated as to be overlooked in normal sampling methods. The inherent limitations on accuracy of analytical methods and the complications of terrestrial weathering further complicate the problem. Masses displaying appreciable weathering are, in fact, of little use for geochemical study. Terrestrial contamination—for example by careless storage—must also be avoided. Some of the early recovered carbonaceous chondrites provide dubious subjects for chemical analyses, on account of contamination during their long sojourn in museum show cases.

Mason recommends the sealing of a portion of any new material recovered after a fall in clean plastic containers: this is especially

---

* Possibly due simply to lesser accuracy of measurement method.

important in the case of carbonaceous chondrites for which new geochemical studies are needed to confirm or refute the results of studies of the older, possibly contaminated, material. The preparation of plaster casts may introduce contamination.

Elemental abundances for all meteorites derived from good chemical analyses are given in Table 11.

TABLE 11

Elemental Abundances in Meteoritic Material (After B. Y. Levin)

Atomic number	Element	ppm Levin *et al.* (1956)	Atomic number	Element	ppm Levin *et al.* (1956)
3	Li	3·2	41	Nb	0·5
4	Be	0·09	42	Mo	5
5	B	2·6	44	Ru	2
8	O	346,000	45	Rh	0·6
9	F	40	46	Pd	0·5
11	Na	7,000	47	Ag	0·5
12	Mg	139,000	48	Cd	2
13	Al	14,000	49	In	0·2
14	Si	178,000	50	Sn	20
15	P	1,600	51	Sb	0·4
16	S	20,000	52	Te	0·14
17	Cl	800	53	I	1
19	K	900	55	Cs	0·08
20	Ca	16,000	56	Ba	7
21	Sc	0·5	57	La	200
22	Ti	700	58	Ce	2
23	V	80	59	Pr	0·8
24	Cr	2,500	60	Nd	3
25	Mn	2,000	62	Sm	1
26	Fe	256,000	63	Eu	0·3
27	Co	900	64	Gd	1·6
28	Ni	14,000	65	Tb	0·5
29	Cu	40	66	Dy	2
30	Zn	20	67	Ho	0·6
31	Ga	8	68	Er	1·7
32	Ge	40	69	Tm	0·3
33	As	70	70	Yb	1·6
34	Se	9	71	Lu	0·5
35	Bi	22	72	Hf	0·8
37	Rb	8	73	Ta	0·3
38	Sr	22	74	W	17
39	Y	5	75	Re	0·0018
40	Zr	90	76	Os	1·1
			77	Ir	0·6
			78	Pt	3
			79	Au	0·26
			80	Hg	0·009
			81	Tl	0·14
			82	Pb	2
			83	Bi	0·16
			90	Th	0·2
			92	U	0·05

Stony and Stony-iron Meteorites

Major element abundances for stony and stony-iron meteorites are given in Table 12 in terms of oxides (after Wood, 1963). In considering the statistics given in this table, the small number of known

TABLE 12

Percentage Compositions of Stony and Stony-iron Meteorites

Meteorite type	No. Averaged	Silicate Fraction															Metal fraction							
		SiO_2	MgO	FeO	Fe_2O_3	Al_2O_3	CaO	Na_2O	K_2O	Cr_2O_3	M	TiO_2	P_2O_5	H_2O	NiO	Total silicates	Fe	Ni	Co	P	Total metal	FeS	C	Ot
Common chondrites	100	38·29	23·93	11·95	—	2·72	1·90	0·90	0·10	0·37	0·	0·11	0·20	0·27	—	81·00	11·65	1·34	0·08	0·05	13·11	5·89	—	—
Bronzite	45	36·41	23·09	8·87	—	2·60	1·87	0·93	0·10	0·33	0·	0·11	0·18	0·30	—	75·05	17·45	1·68	0·10	0·05	19·28	5·67	—	—
Hypersthene	55	39·70	24·58	14·33	—	2·81	1·92	0·94	0·11	0·41	0·	0·11	0·22	0·24	—	85·63	7·13	1·07	0·07	0·04	8·31	6·06	—	—
Cc type 1	3	23·08	15·56	10·32	—	1·77	1·51	0·76	0·07	0·28	0·	0·08	0·27	20·54	1·17	75·60	0·11	0·02	0·00	—	0·13	16·88	3·62	3·
Cc type 2	8	27·31	19·00	20·06	—	2·31	2·03	0·54	0·05	0·39	0·	0·10	0·27	13·23	1·56	87·02	0·00	0·16	0·00	—	0·16	8·58	2·44	1·
Cc type 3	5	33·75	23·86	24·32	—	2·65	2·32	0·55	0·05	0·51	0·	0·12	0·32	1·00	0·33	89·98	2·34	1·08	0·06	—	3·48	6·08	0·46	
CEn	8	38·62	21·01	1·69	—	1·87	0·97	1·00	0·11	0·35	0·	0·06	0·20	0·62	0·00	66·64	19·82	1·66	0·12	—	21·60	10·70	0·29	0·
Amphoterites	3	40·76	26·54	20·17	—	1·81	1·68	1·04	0·24	0·47	0·	—	—	—	—	93·07	1·97	0·90	0·05	—	2·92	4·01	—	—
Eucrites	13	47·60	8·46	15·58	0·98	13·01	10·18	0·43	0·06	0·36	0·	0·43	0·09	0·61	—	98·26	1·18	—	—	—	1·18	0·56	—	—
Shergottites	1	49·92	9·95	18·92	3·15	5·87	10·35	1·27	0·18	—	—	—	—	—	—	100·00	—	—	—	—	—	—	—	—
Howardites	8	49·27	11·76	15·58	1·56	9·95	7·71	1·06	0·36	0·53	0·	0·10	0·08	0·33	—	98·95	0·35	0·10	—	—	0·45	0·60	—	—
Aubrites	4	54·01	35·92	10·97	—	0·67	0·91	1·32	0·10	0·06	0·	0·06	0·22	1·14	0·26	95·78	2·29	0·17	—	—	2·46	1·25	—	0·
Diogenites	4	52·11	25·85	16·05	—	1·18	1·41	0·004	0·001	0·80	0·	0·19	0·01	0·14	—	98·06	0·79	0·03	—	—	0·82	1·12	—	—
Ureilites	2	38·90	35·73	12·73	—	0·38	0·79	0·43	—	0·43	0·	0·09	0·07	1·13	—	91·03	8·13	0·15	—	—	8·28	—	0·69	—
Angrite	1	43·92	10·04	8·28	0·31	8·72	24·50	0·26	0·19	—	—	2·39	0·13	—	—	98·74	—	—	—	—	—	1·26	—	—
Nakhlites	1	48·75	11·96	19·42	1·28	1·73	15·11	0·41	0·14	0·33	0·	0·38	—	0·24	—	99·84	—	—	—	—	—	0·16	—	—
Pallasites	10	17·05	19·83	6·65	—	0·38	0·28	0·07	0·03	0·68	0·	0·00	—	—	0·29	45·34	48·98	4·66	0·30	0·11	54·05	0·53	0·08	—
Siderophyre	1	34·61	10·08	4·40	—	—	—	—	—	—	—	—	—	—	—	49·09	46·02	4·74	0·15	—	50·91	—	—	—
Lodranite	1	28·94	23·33	7·71	—	0·19	0·18	—	—	0·17	—	—	—	—	—	60·52	27·77	4·05	—	—	31·82	7·40	—	0·
Mesosiderite	4	19·51	6·36	5·73	1·95	4·10	2·89	0·17	—	0·36	0·	—	—	0·69	0·40	42·38	45·95	4·39	0·28	0·11	50·73	2·83	—	4·

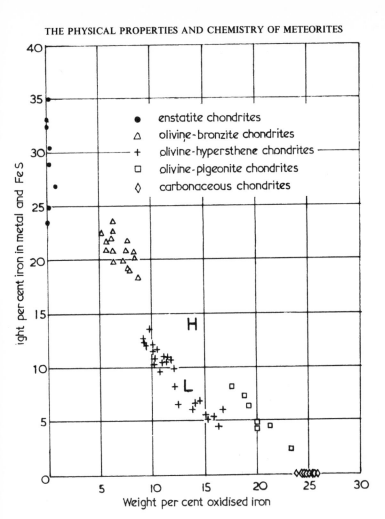

Fig 17 H and L chondrite groups: diagram illustrating their separation by means of a plot of oxidised iron against iron present in metal or sulphide form (from Mason, 1962)

meteorites of certain rare types must be taken into account—one new find can radically change the averages for such groups.

Observations of importance based on studies of meteorite geochemistry include:

155

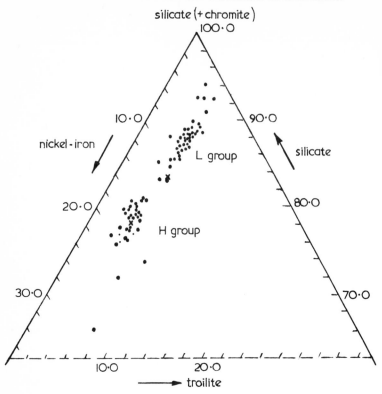

Fig 18 Silicate + chromite, nickel-iron, and troilite content of chondrites plotted on a triangular diagram to illustrate the H and L group partition (after K. Keil and K. Fredrikson)

Nordenskjold (1878): chondrites reveal similarity in their bulk chemistry, but varying oxidation states.

Prior (1916): the less the amount of nickel-iron present in chondrites, the richer in nickel is that metal fraction and the richer in iron are the ferromagnesian silicates.

Urey and Craig (1953): two-thirds of all chondrite analyses were dismissed as unreliable, but from the remaining one-third ('superior analyses') they deduced that there is no continuous spread of oxidation values for common chondrites, which fall into two distinct groups corresponding to the two classes. Also, one of these groups is richer in total iron, nickel, and cobalt computed on an oxygen- and sulphur-free basis (Table 13, from Wood, 1963). Urey and Craig

TABLE 13

Metal/Silcon Ratios for Bronzite and Hypersthene Chondrites
(From Wood, 1963)

	Average CBr	Average CHy		Average CBr	Average CHy
Mg/Si	0·95	0·92	Fe/Si	0·83	0·60
Al/Si	0·084	0·084	Ni/Si	0·047	0·028
Ca/Si	0·055	0·052	Co/Si	0·0028	0·0018

TABLE 14

High and Low Iron Groups (From Wood, 1963)

	Average composition, weight per cent			
	Nickel–iron	Troilite	Chromite	Silicate
L Group	6·54	5·54	0·27	87·65
H Group	16·80	5·03	0·22	77·95

erected high (H) and low (L) iron groups, the meaning of which has been discussed exhaustively by geochemists since. Their reality is undisputed, but their significance remains obscure. As seen in Table 14 (after Wood, 1963), the two groups have markedly different mineral phase ratios.

Wiik (1956) noted that carbonaceous chondrites contain more water, sulphur, and carbon than common chondrites, are more highly oxidised, but display a comparable metal content, and fall within the H group alongside the olivine bronzite chondrites. The three types, 1, 2, and 3, are mineralogically, texturally, and geochemically distinct: ferrous oxide, carbon, and water are the most variable components. He also notes that the enstatite chondrites are, in contrast, highly reduced, though their chemistry is far from uniform: a selection made by him from observed falls eliminated most of this spread of values, and it appears to be a weathering effect imposed after Earth impact. These rare chondrites fall within an extension of the high total iron group of Urey and Craig, and an HH group to cover them is sometimes proposed.

Keil and Fredriksson (1964) established the existence of an extreme low iron group, embracing the amphoterite class of modified and recrystallised chondrites, and some unequilibrated chondrites. These authors have also revised the mineral phase limits of the H and L groups.

Wood (1963) advanced arguments based on J. W. Gibbs's Phase Rule and the fact that the oxidation state of a rock/open atmosphere system is dependent on equilibria relationships with a given atmosphere (which introduces the possibility that meteorite oxidation states preserve a measure of different atmospheric environments). He suggested that either chondrules have remained as closed systems since their formation, or have been open to oxidising or reducing atmospheres: or were neither open nor closed, but had varying amounts of a reducing agent, carbon, lost irretrievably from the system once reduction was effected, the carbon forming carbon dioxide.

Despite the uncertainties concerning the significance of H and L groups and oxidation states, these groups by their existence severely limit theorising on meteorite genesis. Distinct patterns of disparity correspond to mineralogical patterns, but the scale and significance of these patterns in relation to the past history of the meteorite mass remains unknown. We may ask the question: are these patterns indicative of disparity between separate parent bodies or are they simply patterns indicative of contrasting areas in the same parent body? A clue comes down to us in the form of the Bencubbin stony-iron (Chapter 9), in which fragments of enstatite and common chondrite occur as enclaves. These are H and L types respectively, and it is very difficult to accept that these two enclaves stem from different meteorite parent bodies, the evidence being considered in Chapter 13. Presumably, therefore, H and L variants represent (or at least may represent) contrasting areas within a single meteorite parent body: and as these enclaves are highly reduced and highly oxidised respectively, the same must be concluded for contrasting oxidation states.

Wood (1963) noted that chondrites show a marked similarity of chemical composition to the solar atmosphere, except for anomalous iron values. Certain other discrepancies became apparent in his compilation of trace-element abundances, covering all elements except krypton and certain shortlived radioactive elements. His values are based on studies by optical and X-ray spectrography, colorimetry, isotope dilution, and neutron activation methods, of Ehmann, Suess, Urey, Goldschmid, and I. and W. Noddack. They are given in Table 15.

TABLE 15

Trace-element Concentrations

Atomic No.	Element	Chalcophile (C) Lithophile (L) or Siderophile (S)	Concentration atoms/10^6 silicon atoms
2	He	—	0·11
3	Li	L	50
4	Be	L	0·64
5	B	S	40
6	C	S	2,000–20,000
7	N	—	90
9	F	L	300
10	Ne	—	0·0015
17	Cl	L	1,000
18	A	—	0·4
21	Sc	L	30
23	V	L	160–900
27	Co	C, S	1,200
29	Cu	C, S	190
30	Zn	C	120
31	Ga	S	12
32	Ge	C, S	19
33	As	C, S	4·7
34	Se	C	17
35	Br	L	2–32
37	Rb	L	7
38	Sr	L	20
39	Y	L	3·6
40	Zr	L	65
41	Nb	C, L, S	1
42	Mo	C, S	2·5
44	Ru	C, S	1·6
45	Rh	S	0·27
46	Pd	S	1·1
47	Ag	C	0·13
48	Cd	C	0·064
49	In	C	0·0013
50	Sn	C, S	1·1
51	Sb	C, S	0·1
52	Te	C	1
53	I	C, L	0·05
54	Xe	—	0·000007
55	Cs	L	0·12
56	Ba	L	4·0
57	La	L	0·4
58	Ce	L	1·1
59	Pr	L	0·2
60	Nd	L	0·8
62	Sm	L	0·3

TABLE 15—(*contd.*)

Atomic No.	Element	Chalcophile (C) Lithophile (L) or Siderophile (S)	Concentration atoms/10^6 silicon atoms
63	Eu	L	0·1
64	Gd	L	0·4
65	Tb	L	0·06
66	Dy	L	0·3
67	Ho	L	0·08
68	Er	L	0·2
69	Tm	L	0·04
70	Yb	L	0·11
71	Lu	L	0·03
72	Hf	L	1·2
73	Ta	L, S	0·02
74	W	C, L, S	0·12
75	Re	S	0·001
76	Os	C, S	0·6
77	Ir	S	0·37
78	Pt	S	1·3
79	Au	S	0·13
80	Hg	C	0·08
81	Tl	C	0·0007
82	Pb	C	0·15
83	Bi	C	0·002
90	Th	L	0·026
92	U	L	0·0075

Conclusions that may be drawn from the minor element abundances in chondrites are as follows:

1 Minor elements are depleted in relation to solar or cosmic abundances, such depletion contrasting with the major element abundances (calculated from various lines of evidence).
2 Certain elements—indium, antimony, tellurium, iodine—are more abundant in carbonaceous chondrites than in common chondrites, their abundance in these rare meteorites approximating to solar or cosmic abundances.
3 Primordial gases are enriched in carbonaceous and enstatite chondrites, but are virtually absent from the common chondrites.
4 Carbonaceous and enstatite chondrites seem to be closer in composition to the bulk composition of primordial matter than common chondrites, but thallium, lead, bismuth, and mercury have been depleted in some manner not understood.

The Iron Meteorites

The lack of homogeneity of the iron meteorites makes similar trace-element studies less meaningful: for example, troilite nodules may be irregularly spaced several centimetres apart, and the practice of sampling nodule-free material has become standard. Yet this practice does not result in even sampling of the meteorite as a whole, and how does an analyst obtain a representative bulk sample, including nodular material? Even the metal fraction free of nodules is subject to considerable variation. For example, the contrasting areas of alloy in the Sikhot-Alin meteorite (Chapter 13). A meteorite is essentially a random sample of varying size and homogeneity, just like a rock specimen, and seldom can we bulk a large sample and quarter it as we do with rock samples—there usually is little enough material available!

The fact that meteorites are considered as individuals on a statistical basis in geochemical evaluations disregards the above truth. It may be that we should be studying homogeneous areas within individual meteorites more closely, petrographically, mineralogically, and chemically, but such studies are probably equally as difficult to convert into meaningful quantitative statistics as those based on individual meteorite occurrences as a statistical entity. Major element abundances in iron meteorites are given in Table 16.

TABLE 16

Major Element Abundances in Iron Meteorites (After Wood)

Type of meteorite	No. of meteorites averaged	Major elements							
		Fe	Ni	Co	FeS	Cu	P	C	*Misc*
Octahedrites									
Ogg	18	87·96	6·23	0·48	4·69	0·01	0·15	0·22	0·26
Og	34	87·67	7·11	0·52	3·64	0·17	0·17	0·20	0·52
Om	92	88·95	8·08	0·58	1·80	0·03	0·18	0·08	0·30
Of	37	88·94	8·85	0·56	0·65	0·05	0·17	0·60	0·18
Off	10	86·76	11·80	0·62	—	0·11	0·24	0·01	0·46
Hexahedrites	34	92·76	5·56	0·66	—	0·35	0·29	0·19	0·19
Ataxites (nickel rich)	24	79·63	18·90	1·01	—	0·05	0·12	0·10	0·19

It is common practice nowadays to allow an arbitrary 2 per cent of nodular material in iron meteorite analyses, to overcome the sampling difficulty. Even so, these analyses can never be more than good approximations.

Gallium and Germanium

The irons have been studied statistically using comparisons of the gallium and germanium contents with interesting results, notwithstanding the limitation mentioned above. The pioneer work of Goldlberg, Uchiyama, and Brown in 1951 was followed by a number of such studies. A discontinuous pattern of variation emerges separating four distinct groups (Table 17), which are relevant to discussions

TABLE 17

Gallium and Germanium in Iron Meteorites (After Wood, 1963)

Group	Gallium content	Germanium content
I	80–100	300–420
II	40–65	130–230
III	8–24	15–80
IV	1–3	<1

of probable provenance and cooling histories of meteoritic material. Once again the problem of scale arises, however, and the exact meaning of this separation into groups remains obscure. The most popular explanation is that these groups represent at least four dissimilar parent bodies for the iron meteorites, asteroids of different size and therefore of different cooling behaviour. These groups may, however, represent nothing more than hiatal relationships such as are evident in terrestrial igneous rocks, discontinuities in the fractionation process during the cooling of the alloys, products of discontinuities inherent to the physicochemical systems controlling the cooling history of the alloys. One may envisage such discontinuities occurring within a parent body possessing a central alloy core comparable with the Earth's postulated nickel-iron core; however, the parent body may have had no such core, but, instead, a patchy raisin-bread configuration of metal and silicates.

It seems that it would be unwise to regard all such discontinuities in meteorite geochemistry as relating to different parent bodies or even different shells in the parent body interior. We must always bear in mind the type of discontinuity that is produced in continuous liquid-crystal fractionation in igneous rock cooling systems, due to one mineral phase becoming unstable and being abruptly replaced by

162

one or two others, which may not happen to accept a particular component any longer in the crystal lattice. The gallium–germanium discontinuity may perhaps have been given a very special status, irrespective of the fact that meteorite geochemistry reveals more and more such discontinuities as results come in, and they cannot all refer to separate parent bodies, or even separate internal shells.

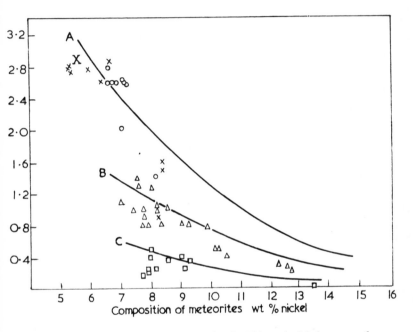

Fig 19 The relationship between kamacite band width and nickel content for meteorites of the four gallium-germanium groups and ten hexahedrites. The curves A, B, and C represent relationships expected to result from cooling rates of approximately 2 °C, 5 °C, and 10 °C per million years (simple model of Yardley for nickel diffusion rates). ○, group I; ×, group II; △, group III; □, group IV. X represents ten hexahedrites (after H. J. Axon)

13: Brecciation in Meteorites

The subject of brecciation in meteorites was brought to the fore by Wahl in 1952. He distinguished *polymict breccias* that carry angular fragments of meteorite material different from that of the host from *monomict breccias* that carry fragments that are no more than pieces of the same material as the host. It is not always certain, however, where the line should be drawn between polymict and monomict breccias: for example, there are breccias that consist of normal chondritic material aggregated with similar material affected by shock and modified thereby. Such breccias are not regarded here as truly polymict, and are discussed separately under the heading of light-dark breccias.

Brecciated Stones

Monomict breccias are commonly developed in achondrite material; by contrast, achondrites only rarely evince polymict breccia character—examples are, however, seen in the Cumberland Falls, Kentucky, enstatite achondrite and the Bencubbin achondritic stony-iron from Western Australia. Chondrites are reported by Wahl to evince polymict breccia character quite frequently, in addition to their almost ubiquitous monomict breccia character. However, a careful study of Wahl's documentation of such polymict breccia occurrences in chondrites reveals that the original attributions were doubtful. Reports of 'aubrite', 'howardite' and other achondrite material forming enclaves in chondrites mainly stem from early workers and so must be treated with reserve, unsupported as they are by modern petrographic or mineralogical evidence, or even by photographs. It is quite easy to confuse large fragments of orthopyroxene derived from chondrules, broken in the process of brecciation, with true achondrite enclaves.

The examples recorded by Wahl of eucrite material set in a howardite

Plate 51 The Cumberland Falls, Kentucky, polymict breccia composed of a white enstatite achondrite host, itself brecciated, and dark recrystallised olivine-hypersthene chondrite enclaves (left). Note the dark fusion crust

matrix are similarly questionable. Chemical and mineralogical definitions nowadays distinguish these two achondrite classes, but in the older classification followed by Wahl all brecciated calcium-rich achondrites were lumped together as howardites, and all non-brecciated varieties as eucrites, irrespective of chemical and mineralogical contrasts. Thus the breccias described by Wahl are probably monomict, consisting of unbrecciated residuals of howardite or eucrite set in a breccia matrix of the same material. The newly discovered Millbillillie eucrite from Wiluna, Western Australia, is reported to be of this type by Binns.

Stony-iron Breccias

The mesosiderite stony-irons are essentially polymict breccias, angular achondrite lumps being set in a microbreccia of silicate minerals of the same character as the enclaves, the metal reticulation bonding the whole together. The polymict breccia of mesosiderites is, however, unusual in that it may involve two or more classificatory types of meteorite, yet the silicate fraction of the host matrix shows some systematic chemical relationship to the metal of the reticulation or the metal which it encloses. In addition, the metal displays a restricted composition range, within the limits of the medium octahedrites—large nodules of metal may in fact display Widmanstätten patterns.

In contrast to this type of polymict brecciation, that of the Bencubbin stony-iron is more complex. The host material consists of a

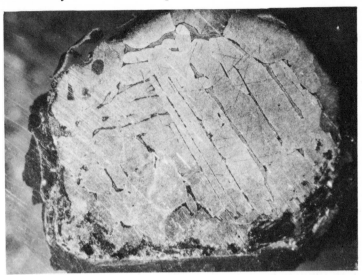

Plate 52 Widmanstätten pattern produced by etching a metal nodule from the Mt Padbury mesosiderite, Western Australia (c × 10)

coarse reticulation of metal of hexahedrite composition, but with no systematic etch pattern, and coarse crystals of enstatite achondrite silicate material fill the spaces in this reticulation: in addition, there are sparse enclaves of diverse types of chondrite material, among which a primitive enstatite chondrite and a primitive olivine-hypersthene chondrite carrying clino-hypersthene, not the usual bronzite, have been recognised. These inclusions reflect a random polymict character such as is quite rare in meteorites. The Cumberland Falls enstatite achondrite, from Kentucky, contains olivine–hypersthene chondrite enclaves and is thus of similar random character.

It is common for meteorites arriving in one multiple fall to be of one classificatory type, but recent discoveries in Western Australia are suggesting that this is by no means always the case. The Dingo Pup Donga ureilite and unusual Coorara chondrite carrying ringwoodite and majorite were found within yards of each other, way out on the wastes of the Nullarbor Plain. The possibility that they fell

167

together is being further investigated; it seems possible that large masses of stony meteorite material that break up into shower fragments as they suffer ablation may well, in rare cases, contain sizeable areas of contrasting classificatory types.

Brecciated Irons

Polymict breccias are not unknown amongst the iron meteorites, though they are rare. Examples include the coarsest octahedrite (or brecciated, 'granular' hexahedrite) Sikhot-Alin, reported to contain granular hexahedrite enclaves in coarsest octahedrite material. However, there is much argument about the true definition of this material —Russian authorities are regarding it as all of coarsest octahedrite material according to Kvasha—and this seems to be more a case of juxtaposition of contrasting fields of metal alloy than true polymict brecciation. The New Baltimore, Pennsylvania, iron provides a much better example, for this meteorite displays an enclave of octahedrite material showing a Widmanstätten pattern within a matrix of large kamacite grains, wholly hexahedritic in microstructure according to Perry, and classified by him as a granular hexahedrite: however, even this example is, again, a case of contrasting metallurgical fields in juxtaposition. Henderson has found it hard to explain iron meteorites that display two types of metallurgical texture in juxtaposition. He does not dismiss shock accumulation on meteoroid collision, but regards the lack of deformation due to such shock on collision as virtually ruling out this explanation. This problem is the same as that involved in rational explanation of such polymict breccia associations as those of the Bencubbin and Mt Padbury, Western Australia, meteorites. It remains doubtful whether there are any true polymict accumulation breccias of iron within iron among the known iron meteorites.

Simple cataclastic brecciation is not uncommon in iron meteorites, as in the case of stones and stony-irons. In particular, irons of the coarsest octahedrite or granular hexahedrite classes tend to come apart and be distorted by virtue of movement along the coarse, intergranular sutures. Late invasions of troilite or schreibersite may enter such sutures and ramifying systems of cracks traversing the face of the iron meteorite. There may, however, be no such sulphide or phosphide introduction, the only evidence of brecciation being the distortion of Neumann lines in kamacite grains close to these sutures and cracks.

Plate 53 A faceted eucrite enclave separated from the metal of the Mt Padbury mesosiderite along the boundary represented by the facets (7·5 cm diameter)

Coming to the term 'granular', it is a fact that, as in geology, the terms granular and granulation have been overworked. The so-called granular hexahedrites and coarsest octahedrites that display the same coarse granularity grading into irregular Widmanstätten pattern are commonly regarded as brecciated irons, but in fact only display varied orientations of Neumann lines in the kamacite through the granular aggregate where there has been cataclastic deformation along sutures and cracks—the granularity itself is a metallurgical effect, not a brecciation effect. Hexahedrites are known to consist of single crystals of great extent—whole masses show the same crystallographic orientation—and the granular hexahedrites are products of metallurgical granulation developed within the crystal.

Quite different is the 'granularity' of the so-called granular octahedrites, exemplified by the Four Corners, New Mexico, iron and the newly discovered Mundrabilla iron from Western Australia. These meteorites consist of areas of diversely oriented octahedrite material, each area having its own particular Widmanstätten pattern orienta-

169

tion. Such areas may be defined by tight interfaces between alloy areas, by irregular areas of troilite, schreibersite, and graphite, or may even possibly be islanded by such areas. The characteristic of the Mundrabilla iron, to shed many small iron masses for many miles around the site of landing of the two main masses (broken from one mass just prior to impact), may well be due to the weakness imposed by the many areas of troilite, schreibersite, and graphite situated between the alloy areas. In addition, some of these meteorites contain small patches of silicates—olivine, orthopyroxene, and calcic plagioclase. Mundrabilla and Four Corners do, the former containing olivine and orthopyroxene. This feature requires the classification 'granular octahedrites with silicate inclusions'. This characteristic appears to be an incipient development of stony-iron character, with the metal fraction still overwhelmingly dominant.

Plate 54 Cut, etched, and polished surface of the Mundrabilla 'granular' octahedrite, from Western Australia—one of the small, shed masses. The varied orientation of areas of Widmanstätten pattern is evident, also large graphite areas (dark) rimmed with sulphide (light)

Irons with Silicate Inclusions

This incipient development is taken a stage further by the example of the Woodbine, Illinois, meteorite, classified by Mason as an iron with silicate inclusions. It contains more silicate material within the octahedrite host, which shows a well-defined etch pattern, but still not enough to be strictly classified as a stony-iron. It is a transitional type between the Mundrabilla and Four Corners type and the true stony-irons. Besides its metal–silicate ratio, the nature of the silicate inclusions is anomalous—Mason shows that it has chondritic chemistry: at present, nothing has been published on the texture of the

Plate 55 Cut, etched, and polished surface of the Woodbine, Illinois, meteorite showing dark angular enclaves of chondritic composition, set in a metal alloy that contains finer silicate enclaves and displays a medium-coarse Widmanstätten pattern

included material, but, as far as is known, there is no preservation of actual chondrules and the material must be regarded as highly modified (? recrystallised) chondritic material. If this is correct, the state of thermal modification contrasts with the quite unmodified state of the enclaves of chondritic material in the Bencubbin meteorite.

The meteorite Nyetschaevo, from the USSR, is an octahedrite with chondrite inclusions (pers comm, L. Kvasha), and the author found a small sample that he examined in Moscow very similar to areas in the Woodbine meteorite, though the slice was too limited in area to show the overall structure well. The inclusions are reported to be chondritic, and Olsen and Jaresowich (1971) illustrate actual fan and monosomatic chondrules. The chondrite inclusions are bronzite and enstatite chondrites of chemistry and mineralogy intermediate between them.

These two transitional meteorites should perhaps be grouped with Mt Egerton as transitional stony-irons. The class could be divided into two sub-classes with the two octahedrites containing chondritic material in one, and the essentially achondritic meteorite in the other.

The stony-irons with chondrite inclusions (represented by the two Bencubbin masses only, at present) are probably end-members of the sequence: 'granular' irons with minor amounts of silicate inclusions (Mundrabilla, Four Corners), transitional irons (Woodbine, Nyet-schaevo), stony-irons with chondritic inclusions (Bencubbin) and stony-irons with achondritic inclusions (Mt Padbury); transitional stony-irons consisting of achondritic material with metal alloy inclu-sions (Mt Egerton); chondritic and achondritic stones. This series illustrates a progressive change in metal–silicate ratios only: it ignores distinct petrographic and chemical groupings.

This type of polymict association is not a product of cataclasis. These are accumulation breccia associations, one type of material being aggregated within another to varying extents, possibly quite irrespective of deformation. The 'granular' octahedrites taken on their own, however, do not appear to be accumulation breccias. The irons with silicate inclusions may be regarded as the incipient accumu-lation breccias, but only if one accepts that the accumulation of one type of material in another may have taken place very early, in the parent body, and may simply be due to a partition of immiscible melt fractions. These inclusions may be entirely crystalline or part glass, the latter indicating that they have been at least in part molten during their history. They commonly contain carbon and this points to the close genetic connection between troilite–schreibersite–graphite segregations in irons and silicate inclusions. The iron meteorite is predominantly a product of crystallisation of Gold-schmidt's siderophile phase, and minor residues of the other two im-miscible phases—chalcophile and lithophile—are represented by troilite–schreibersite–graphite areas, and by silicate inclusions respectively. It is not chance that produces the silicate inclusions in the granular octahedrites containing unusual quantities of troilite etc: these are meteorites containing residual amounts of both the other two immiscible phases: the chalcophile phase in fact goes along with virtually all types of meteoritic segregation.

Allocation of Brecciation to the Correct Stage in the History of the Meteoritic Mass

One of the major problems of meteoritics is to determine at what stage in the history of the meteoritic material a particular brecciation effect occurred, whether it be a simple cataclastic effect or an accumulation effect. There are four alternative possibilities:

1 *In the parent body:* by internal disruptive or gravitational mixing processes or simply by heterogeneous arrangement of immiscible liquid phases: the heterogeneity could even be inherited from an original heteorogeneity in the post-accretion melted solid accretion or spray accretion product (for example a raisin-bread configuration of diverse lumps swept up from the disseminated material of the nebula). In this case the breccia would only be an accumulation breccia in the broadest sense, being formed by melting and congelation of a heterogeneous mass in situ, without any transportation and introduction of exotic material, even from other parts of the parent body.

2 *In the post-disruption or meteoroid stage* due to collisions, aggregation together, or shock deformation after the separation of discrete individual meteoroids from the parent body, and while they were orbiting around the Sun, individually.

3 *In the atmospheric entry stage*, due to shock consequent on meteorite ablation buffeting with or without fragmentation.

4 *In the impact stage* due to shock on impact with the surface of the Earth.

Taking the impact stage (4) first, such effects are of minor significance in modifying meteorite masses, unless the mass is of very large dimensions, and thus impacts at unabated cosmic velocity (Chapter 20). Individual fusion-crust coated masses may suffer fragmentation, forming masses with fresh, broken, uncoated surfaces or fragments traversed by unfilled, partly-opened cracks. Such fragmentation detritus will be scattered around the point of impact in a litter of boulders, probably showing some systematic distribution. No thermal

173

(a)   (b)

(c)   (d)

Plate 56 *Achondrite enclaves in the Mt Padbury mesosiderite: photomicrographs.*
(a) Eucrite consisting of pigeonite and lamellar twinned bytownite, showing granu-
litic texture; (b) Eucrite consisting of pigeonite (grey) and bytownite (white: droplet
inclusions), together with tridymite (white); (c) Eucrite showing pigeonite rimmed by
hypersthene and displaying exsolution lamellae, a characteristic of slow-cooled
magmatic rocks; (d) Eucrite showing ophitic intergrowth of narrow plagioclase
laths and pigeonite, a texture typical of terrestrial dolerites; (e) Crystal breccia
texture in a hypersthene achondrite enclave. Orthopyroxene, grey; tridymite, white,
interstitial; (f) Crystal breccia texture in an olivine achondrite enclave; (g) Dark,
narrow 'flinty crush' veins seen under high magnification in the same enclave; (h)
Cold granulation texture in an eucrite enclave. The ophitic texture of the pigeonite
(dark) and plagioclase (white) intergrowth is preserved in vague outline, despite the
recrystallisation to form a fine granular aggregate. The possibility that this is the
incipient stage of the process that formed the maskelynite areas in the Shergotty,
India, shergottite must be entertained. (Magnifications c × 25 and × 63)

174

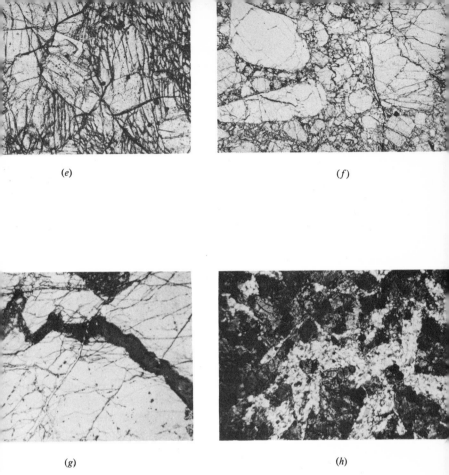

(e)

(f)

(g)

(h)

modification normally accompanies such brecciation. Iron masses may leave serrated surfaces free of ablation marking exposed to the air; surfaces which may, after terrestrial weathering, display the crystallographic planes even before the meteorite is etched.

The third process, brecciation during ablation flight, either high in the Earth's atmosphere or relatively low down, is a more important process. It is this process that is responsible for the fall of meteorites in showers of masses, for most entries into the atmosphere are initially of single masses. Processes acting are shock effects consequent on deceleration from cosmic velocity (akin to the well-known sound-barrier buffeting experienced by supersonic jet aircraft), and abrupt fragmentation in an explosive manner later in the downwards passage of the bolide, the observed expansion of the luminous globe due to sudden release of a gas phase, possibly from cracks traversing the mass. The bolide may either swell suddenly or even split into several parts. Violent release in pressure must accompany such outbursts, whatever their origin. The product from this or buffeting disintegration is a mass of individual stones or irons, or stony-irons. If the fragmentation occurs early enough in ablation flight, a new fusion crust may partly cover the fresh broken surfaces exposed to ablation, and we get the sort of contrast of surfaces well exemplified by the members of the Wiluna, Western Australia, shower of stones in 1967. Krinov has classified these ablation surfaces as 'surfaces of the first and second kind'. At the same time as fragmentation occurs, other parts of the masses may only be penetrated by breccia veins of angular fragments within dark, magnetite-rich glass or clear glass, or veins simply of glass: such veins will run out to merge with the fusion crust when seen in a section cut across the interior of the mass and fusion crust, and this relationship distinguishes them from early formed veins formed prior to the entry of the mass into the Earth's atmosphere: such veins will be truncated by the fusion crust abruptly. Veins of breccia or glass formed during ablation are actually unopened surfaces of Krinov's second kind: fragmentation was never completed on these particular surfaces.

In some cases dark veins of this kind found in stony meteorites are of doubtful origin: that is, there is no evidence to indicate at what stage they were formed, because they do not appear in contact with the fusion crust surface. Of particular interest are the veins of dark glass carrying the dense spinel-form polymorph of olivine, ringwoodite, and the dense garnet-form polymorph of pyroxene, majorite, noted in both the Coorara, Western Australia, and Tenham, Queensland, common chondrites. These are produced by a high-pressure transformation of the normal silicates, and are the mineral phases predicted to occur within the second layer of the Earth's main silicate shell or mantle. This was their first recognition in Nature. At present it is not known at what stage in the history of this material this transformation occurred. Whereas the possibility that it occurred in the parent body stage must be entertained, and production during collisions in the meteoroid stage cannot be ruled out, it is noteworthy that whereas Coorara is a single isolated find, Tenham is a prolific multiple recovery, and the Tenham masses are thus likely to show effects of ablation-stage brecciation. The suspicion that this is indeed the stage to which such transformations must be referred is strengthened by the recognition of similar veins in masses belonging to the prolific Wiluna 1967 shower of meteorites in Western Australia, though so far no high-pressure phases have been identified, the material not having been examined with suitable equipment.

Brecciation by virtue of aggregation of masses or collision shock in the meteoroid stage (2) is the least well documented of the four possible processes. Whereas most authorities regard this as a major process in the development of meteorites as we see them on Earth, and indeed explain anomalies in cosmic-ray exposure ages of chondrites by invoking such collisions (Chapter 18), spallation producing progressively smaller masses and so offering new areas again and again to cosmic-ray exposure, the breccia textures attributable to this process have yet to be defined. This process is certainly likely to have caused many modifications in the friable chondrite masses and even more friable enstatite achondrites, but may well have been overemphasised to explain experimental anomalies.

Attributions of polymict iron and stony-iron breccia assemblages to collision accumulation in this stage are difficult to accept. Against such attributions must be ranged the lack of evidence of shock in the host or enclave, the restriction of composition of the mineral phases in host and enclave evident in most cases, the congenetic relationship between silicate and metal reticulation in the achondrite association of the Bencubbin meteorite, the host material to the chondrite enclaves being essentially a metal-rich enstatite-achondrite obeying Prior's Rules. In general, in the case of the stony-irons, except in the case of chondritic enclaves, the metal and host in the reticulation are systematically related, and even achondrite enclaves show a systematic relationship to the host reticulation. It is unlikely that meteoroids in Space would continuously collide with related types, though it is not beyond the bounds of possibility, for there is much evidence accruing which suggests that specific petrographic types possess specific orbital characteristics in the meteoroid stage. It is the lack of shock effects rather than the limitation of the mineralogy and chemistry that provides the most cogent argument against meteoroid-stage aggregation on collision being a major factor in producing such associations.

In addition, however, there is evidence that the introduction of enclaves in stony-iron polymict assemblages preceded the cooling of the iron fraction, and therefore must have occurred in the parent body stage prior to disruption. Either the enclaves were moved into place at that time or they are products of accretion even before melting as suggested above. The facts are overwhelmingly against meteoroid-stage collisions being a major agency in producing polymict breccia associations in meteorites. If one wishes to invoke such an aggregation one must surely invoke pre-parent body accretion collisions of planetesimals, and then one comes up against the problem of how far back one can extrapolate the orbital existence of meteoroids in the solar system, that is the problem of whether there was a pre-asteroidal meteoroid population. Shock effects on collision of meteoroids after parent body break-up may well, however, have produced much of the deformational or cataclastic brecciation in meteorites.

Here we must digress and mention the idea that mesosiderites are composite meteorites produced by spallation of lunar eucrite or howardite material by impacting iron meteorites. Popular in America for some years, if somewhat facile, this idea suffers from the objection that mesosiderites are characterised by an octahedrite iron component of restricted composition range, so only these meteorites apparently spalled silicate material off the Moon. Apollo 11 and 12 recoveries have shown that the lunar surface material is quite unlike the eucrite and howardite material even in the nature of the crystalline masses supposedly volcanically excavated from within the Moon's interior: especially noteworthy are the contrasting trace element spectra. Where is the high titanium content of lunar rocks in the mesosiderite which, like all meteoritic material, is titanium-poor? This idea must be relegated to the limbo of interesting concepts which have subsequently proved untenable.

Coming to the parent-body stage (4), we must, it seems, refer polymict brecciation to this stage. There was undoubtedly much stress including violent deformation and shock effects in this stage, as on Earth, and many of the shock and deformation effects seen in meteorites must be attributed to this stage, too.

The simple monomict breccias of the achondrites—enstatite and hypersthene achondrites, and howardites–eucrites—are attributed to deformation in the parent body, in areas of internal stress. The coarse crystalline aggregates of the first two types are particularly susceptible to such deformation, and indeed few examples are known that are not brecciated. The hypersthene-achondrite enclaves in the Mt Padbury achondrite show similar brecciation textures to discrete meteorites of this class, and are bonded by a reticulation of metal alloy which shows little obvious sign of deformation: all the evidence suggests that the brecciation preceded the metal phase invasion or solidification, and that it must be referred to the parent body stage. The same argument shows the cold granulation, similar to 'flaser structure' in gabbros, displayed by the eucritic enclaves in the same mesosiderite to be referable to the same stage. Such granulation has been seen by the author in a discrete eucrite, the partly granulated Millbillillie eucrite. Likewise, the brecciation shown by the olivine-achondrite enclaves in the same meteorite, accompanied by dark 'flinty-crush' veinlets, is referable to the parent-body stage.

Brecciation effects of a purely cataclastic nature displayed by the granular hexahedrites or coarsest octahedrites (Gosnells and Redfields, Western Australia)—the opening up of intergranular sutures and cracks, with or without filling by troilite etc—seem likely to be effects of the late parent body stage, but could represent meteoroid stage deformation. The deformation seen in the Zacatecas iron from Mexico is of similar type.

The Russian authority, Vdovykin, has attributed to shock resultant on meteoroid collision in Space the formation of diamonds in ureilites, and the recognition of the polymorph lonsdaleite, experimentally synthesised by shock, in these meteorites, seems to establish that they are shock products. There seems to be no valid reason, however, for excluding shock in the parent body stage—or on ablation or earth impact for that matter.

The angular enclaves in the Mt Padbury mesosiderite, being of three different achondrite types, provide important restricting evidence. They have faceted form, just like meteorites devoid of fusion

179

crust. This does not mean that they are meteoroids accreted in the mass after parent body break up—indeed the evidence is, as we have seen above, against this. Rather, we may be seeing the embryonic separation of faceted meteorite masses in the parent body, preserved in 'fossil' form. Disruptive effects that are evident within these enclaves may have been precursors of even greater disturbances that eventually broke the parent body into fragments hurled as meteoroids into eccentric, Mars-passing orbits that allowed the possibility of eventual chance tangling with that of our own planet. This structure, of faceted enclaves, is seen in areas of the parent body that did not come apart in the eventual disruption event.

One of the most inexplicable features of meteorite breccias is the recognition of quite primitive, fragile material, liable to modification on any appreciable thermal contact, within high temperature achondrite–metal host assemblages clearly solidified from a melt. The enstatite chondrite/common chondrite enclaves in the Bencubbin stony-iron seem quite inexplicable, for shock effects seem to be quite absent, precluding effects of meteoroid collision. The raisin-like distribution of sparse enclaves in the host material is in any case quite incompatible with such an explanation. If this type of polymict breccia reflects inhomogeneities in the parent body as originally accreted from planetesimals or as it was after internal disturbance, we still have to explain the lack of shock effects in the host material and the lack of thermal modification of the enclave material. The high-temperature character of the host material and primitive nature of the enclaves that have apparently survived its congelation are difficult to explain in terms of any theory.

The enclaves are of type 3 in the classification of Van Schmus and Wood (Chapter 19), showing no signs at all of thermal modification (recrystallisation), and preserving the glass in the chondrules. They are among the most primitive chondritic material ever recovered, and the enstatite chondrite has only very poorly formed and incipient chondrules. A newly discovered amphoterite enclave is similar.

Even more strange is the case of the Holyoke, Colorado, common chondrite which contains a small area of carbonaceous chondrite type 2 material set in the chondritic host (Ramdohr, pers comm). Again, how could such material survive the thermal conditions of crystallisation of the envelope, unmodified and preserving its knife-sharp contact?

Such inclusions must be regarded as xenoliths or autoliths caught up in some way in the crystallisation of the host, and in all probability in the congelation of a melt or spray. They do not seem to be in equilibrium with the host material, but neither do the calcium-rich and calcium-poor achondrite enclaves in the Mt Padbury mesosiderite, for these include silica-undersaturated olivinic and silica-oversaturated tridymitic varieties! It is doubtful if these can have been in equilibrium with the host or each other.

The suspicion arises that the peculiar physical conditions within the interior of the small asteroidal parent body allowed such patches

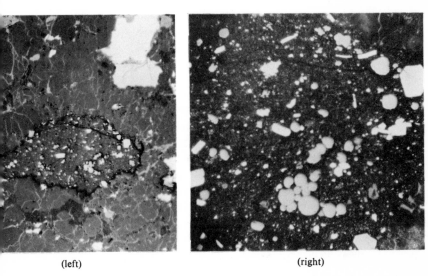

(left)

(right)

Plate 57 Polished section of a carbonaceous chondrite type 2 enclave in the recrystallised olivine-bronzite chondrite from Holyoke, Colorado. Left half, under c × 200 magnification, showing the knife-sharp interface between enclave and host; right half, under c × 650 magnification, showing reflectant spheres of pyrrhotite and troilite set in a serpentinous base

to remain unequilibrated with the molten host or spray-like host and with each other, and also to suffer little or no thermal modification by the host. There may be an analogy with the inhomogeneity of the upper mantle indicated by kimberlites.

It remains possible that volatiles were held in such areas and that they and the carbon content exerted some inhibiting influence on reaction. We need not assume that volatiles will always have escaped when the area was heated above a certain critical temperature, for the envelope may well have exerted an encapsulating effect.

181

A related occurrence was reported from the Leoville, Kansas, carbonaceous chondrite of type 3, which contains enclaves of type 2. The report of achondrite fragments in the same occurrence is, however, misleading: these fragments are of the now familiar spinel-perovskite-gehlenite-anorthite type found in the Allende carbonaceous chondrite of type 3 (Chapter 11). They are thus products of devitrification, not exotic material or material of a contrasting meteorite type.

'Tuffaceous' Textures

Chondrites are commonly referred to in the literature as tuffaceous, on account of a supposed resemblance to terrestrial fragmental volcanics (pyroclasts), but, though they do commonly consist of chondrules, broken fragments of chondrules, and individual crystals of the chondrule-forming minerals set in a base of finely broken chondrule material, crystals, and possibly glass, there is no close resemblance to the textures of shardy volcanic tuffs. The so-called tuff textures are in reality crystal microbreccia or crystal-glass microbreccia textures, and the resemblance, if any, is to crystal tuffs. The chondrites do, of course, show coarser breccia structures already described, but these have nothing to do with the supposed tuff analogy, which depends on the essential microbreccia texture. This microbreccia character must be referred, at least in part, to the parent body stage—for example, the Bencubbin chondritic enclaves show this texture, apparently already formed prior to congelation of the host metal reticulation.

Light-dark Breccias

Another type of brecciation seen both in chondrites and achondrites is the strange light-dark pattern. This is not nowadays regarded as polymictic brecciation, the dark portions being simply modifications of the lighter material involving depletion due to shock of certain elements. A very careful study by Fredriksson and Keil of the Pantar, Philippine Islands, and Kapoeta, Sudan, stones of this type (an olivine-bronzite chondrite and a howardite achondrite) has shown that:

1 The mineral phases in light and dark fractions are the same.
2 The darker fraction has the finer grain size.
3 Single crystals in the light fraction may project into the dark fraction.

4 The light areas tend to be surrounded by the dark.

5 These are monomict shock breccias.

6 There is an anomalous concentration of primordial noble gases in the dark fraction, attributed tentatively to selective trapping in some recrystallised minerals of the dark fraction, more severely altered by shock waves.

The presence of large amounts (relatively speaking) of primordial gases in the dark fraction suggests that the shock occurred within the parent body, early in the history of the meteoritic material and before parent body break-up.

Summary of Brecciation in Meteorites

In conclusion, a summary is given here of the most acceptable interpretation of meteorite brecciation phenomena in the light of the evidence:

1 Accretion of lumpy material of diverse character from the solar nebula formed small parent bodies with raisin-bread structure, metal and contrasting silicate fractions being patchily distributed rather than distributed by the agency of gravitation into well-defined, discrete shells. This patchiness may contrast with the interior of the Earth, which possessed a higher gravitational field, but of course it may well be that similar heterogeneity exists deep within our own planet. Achondritic material may represent material from a deeper level of accretion than chondritic material as a general rule, hence its more common accretion in the denser metal phase which will, on account of its denser nature, have tended to gravitate to the centre. The fact that chondritic material has been found within iron meteorites and stony-irons means that it did, on occasion, somehow find its way to the deeper levels, though it mostly represents material from the outer and more dispersed envelope, crystallised during the degassing phase, perhaps from a fog or spray in most cases.

2 Accretion was followed by heating-up of the parent-body interior, and then by cooling, and there must have been much disturbance and deformation of a violent nature accompanying these stages, prior to parent-body disruption—however, though some of the inhomogeneities, some of the polymict associations, some of the cataclastic effects and some of the deformation and shock effects must be attributed to this disturbance, some of the heterogeneity

183

was probably due to inherited inhomogeneities surviving from accretion, or separation of immiscible melt phases and intermingling of them during internal melting. Explosive degassing may have been involved in these internal disruptions, and it may have reached a climax to effect the final disintegration of the parent body.

3 Very few of the brecciation textures that we see in meteorites are considered to be due to meteoroid collision with resultant aggregation or deformation in the meteoroid orbiting stage.

4 Some brecciation effects involving glass veining may be attributed to shock resultant on sound-barrier buffeting and internal phase-change expansion effects of explosive nature during atmospheric entry.

5 Impact on the surface of the Earth mainly breaks up meteorite masses, except in the case of very large masses, which may, together with the impacted surface rocks, suffer significant brecciation and melting (Chapter 20).

6 Terrestrial agencies finally break up and destroy meteorite masses: however, such effects are easily distinguishable from cosmic brecciation and ablation brecciation.

14: Carbonaceous Chondrites—
Primitive or Degraded?

As has been noted in Chapter 11, these rare meteorites are possible oxidation products of other types of chondrite material, but with inert gas and bismuth-thallium-lead-mercury trace element group contents anomalously high. However, discussion has ranged widely over the problem of the carbonaceous chondrites during the last decade, and whereas certain authorities regard the type 1 carbonaceous chondrites as primitive and parental to all other more reduced chondrite types, others regard them as products of secondary and retrograde alteration of high-temperature types, by processes involving oxidation and hydration; and more recently evidence has suggested that both types 1 and 2 carbonaceous chondrites are mechanical mixtures of material formed in widely different temperature environments.

It is customary to assume that because chondrites do show apparent progressive chemical and mineralogical changes throughout the classes (Cc1 and 2, Cc3, CHy, CBr, and CEn) one of these classes *must* be a common parent to the others. This may, however, be an incorrect assumption. It may be that the chondrites have been to a too great extent separated from the other classes of meteorites in genetic discussion, and that such a separation is unrealistic. If, however, *any* of the five classes of chondrites does have a common parent status, it is the carbonaceous chondrites of type 1 that are the only possible candidates for such a status.

Numbering no more than a score in all, and all these recovered after observed falls, the carbonaceous chondrites of types 1 and 2 comprise no more than a very small fraction of the total meteorite representation. This scarcity in itself is perhaps against a common parent role for either both types or type 1 alone. Of the three types of carbonaceous chondrites distinguished by Wiik, only the type 1 variety was then considered to contain no minerals other than hydrated, low-temperature mineral species, and even this has since

185

become open to doubt, for both Mueller and Kerridge have reported olivine from type 1 carbonaceous chondrites. Type 2 carbonaceous chondrites certainly contain the high-temperature silicate minerals olivine and enstatite, both characteristic of crystallisation from high-temperature silicate melts, and in addition they contain the low-temperature hydrous silicate mineral species predominant in type 1. The two high-temperature silicate minerals are those commonly supposed to form the bulk of the upper layer of the earth's main silicate layer or 'mantle', and are quite anhydrous.

In contrast, serpentine, a sheet mineral family closely related to the chlorite family of sheet silicate minerals, is a low-temperature mineral, and is hydrous. It is not a mineral characteristically crystallised from igneous melts in the magmatic stage, but is characteristically the product of deuteric hydrothermal alteration ('stewing in their own juice') of igneous olivines and magnesian pyroxenes, such as enstatite and bronzite. Chlorite has also been recorded from carbonaceous chondrites, but the hydrous sheet mineral present is usually referred to as serpentine—the definition of the two mineral groups is in any case somewhat controversial. Both Russian and American scientists have conclusively demonstrated the breakdown of meteoritic olivine to serpentine in some stony meteorites, by virtue of textural relationships indicating modification, seen microscopically. In the case of the serpentine mineral known as 'Murray F mineral', present in carbonaceous chondrites of type 2, Anders records that the mineral is randomly oriented, whereas islands of olivine display common optical orientations, a normal feature of olivine residuals left as island relics of the original olivine crystal grain, amid a pool of disoriented secondary serpentine in terrestrial igneous peridotite rocks. This evidence surely favours a degraded status for type 2 carbonaceous chondrites and the evidence of Mueller and Kerridge referred to above suggests that type 1 carbonaceous chondrites may too be of degraded origin, the hydration and oxidation being a reflection of the degradation process.

Against this evidence has been the fact that certain trace elements are present in their cosmic abundance in these meteorites, but not in other types of chondrite. It is argued that these trace elements could not conceivably have suffered differential depletion, and then subsequent restoration to reinstate the cosmic abundances. Some scientists, however, remark that the depletion noted in other chondrites may very well have occurred, the trace elements being lost as con-

186

centrates in water and carbon geochemically removed from the system, but even then, the carbonaceous chondrites, though not so depleted, might still *not* be common parental material to the other chondrite types.

Ringwood, like Mason, favours the common parent status, but only regards type 1 as having this status, so avoiding the difficulty of high-temperature mineral phases. Recent evidence suggests, however, that this limitation does not eliminate this objection, for it appears that no type of carbonaceous chondrite is free from high-temperature mineral phases. Ringwood's argument is that the other chondrites could only be derived by removal of certain trace elements from the material of type 1 carbonaceous chondrites. He cites evidence that the type 1 carbonaceous chondrites have had a far simpler chemical history than any other chondrites and notes that their composition is closest to primordial elemental abundances derived from nucleosynthesis inferences, to solar abundances obtained spectrometrically, and to the composition of the solar atmosphere similarly derived. He suggests that the high oxidation state and volatile content of these meteorites indicates a close relationship to the primordial dust of the parental solar nebula. He regards type 2 as a physical mixture of primitive material of the same sort and high-temperature minerals. It seems that all carbonaceous chondrites must now be so regarded.

Nagy notes the fact that the magnesium sulphate in type 1 carbonaceous chondrites occurs in veinlet form: such a form seems only compatible with hydrothermal introduction during alteration under low-temperature conditions. Studies of the Orgueil meteorite, of this type, have in fact indicated a series of successive stages in mineralogical development. The work of Bostrom and Fredriksson, based on phase equilibria calculations, indicates three stages in development:

1 Early stage—several hundred degrees Celsius.
2 Middle stage—below 170° C.
3 Final stage—below 50° C.

Oxygen isotope determinations by Taylor and co-workers indicate that these meteorites underwent a series of changing environments of crystallisation. Pseudomorphs of what appears to have been anhydrous, high-temperature silicates, replaced by chlorite, are also referred to by Nagy, who suggests that the textures of these meteorites seem to indicate a history of repeated brecciation and hydrothermal alteration, with recementation and leaching by aqueous solutions.

The anomaly of the primitive composition, so similar to the un-fractionated gas cloud of the nebula, remains: yet the truth is that the type 1 carbonaceous chondrites must have suffered considerable chemical modification, for veining by magnesium sulphate must surely effect just such modification? Magnesium, sulphur, and oxygen *must* have been added and there *must* have been associated leaching, modifying the meteorite bulk composition. Nagy makes a very good case against a primitive unmodified status for the carbo-naceous chondrites of type 1: in doing so he concludes that there is as yet no solution to the problem of how a meteorite mass can be chemically fractionated by alteration and leaching processes, and yet have bulk chemistry little different to that of the unfractionated gas cloud of the nebula. The argument of Ringwood is that a two-fold process of fractionation depletion and subsequent restoration is un-likely in the extreme: Nagy seems to suggest that, unlikely and un-explained as it is, it seems to be the only valid explanation of the anomalies.

The contrast between the mineral species of the carbonaceous chondrites of type 1 and 2 introduces another anomaly, for in type 2 they are the highly magnesian high-temperature species of the iso-morphous series which have magnesian and ferran end members. Never do we find the olivines and pyroxenes of the common chond-rites, which contain mineral species containing 20–30 per cent of the Fe ion, though still predominantly magnesian. In contrast, the olivines of the carbonaceous chondrites of type 1, although only detected in trace quantities, appear to be highly ferran, and closely resemble the species occurring in type 3 carbonaceous chondrites ('olivine pigeonite chondrites'). Thus we have an indication that type 2 carbonaceous chondrites are more closely related to the highly reduced enstatite chondrites despite the fact that those are the only chondrites lacking olivine, and that the type 1 carbonaceous chond-rites may be more closely related to the highly oxidised carbonaceous chondrites of type 3. Thus types 1 and 2 are apparently not closely related genetically, whereas 1 and 3 are, despite the fact that physic-ally types 1 and 2 most closely resemble one another, and type 3 more closely resembles the common chondrites! The conclusion that carbonaceous chondrites of type 1 and 2 are highly degraded mech-anical mixtures or superimpositions of high- and low-temperature

188

materials, and also include carbonaceous representatives of widely different and hiatally separated genetic classes, that have come to resemble one another more by virtue of a similar late-stage history of modification producing their carbonaceous nature and volatile enrichment, seems inescapable. They seem to have had no early stage consanguinity. It seems that two diverse genetic groups at different ends of the chondrite range have, by some process not fully understood, both undergone some form of hydration, carbonation, and oxidation at low temperatures accompanied by brecciation, and by virtue of a form of convergence, have come not only physically to resemble one another, but also to display very similar trace elements.

Geochemistry

Mention has been made in Chapter 13 of the discovery by Ramdohr of type 2 carbonaceous chondrite material as an enclave in the Holyoke recrystallised olivine bronzite chondrite. It is difficult to comprehend how such a fragment, containing serpentine, magnetite spherules, and ideomorphic pyrrhotite grains, could have survived the strong recrystallisation of the enclosing high temperature chondritic material, yet it is difficult also to equate such an inclusion with any purely mechanical process of mixture, such as late shock brecciation consequent on collision of two contrasting meteorites after separation from their parent body or bodies. This extraordinary occurrence suggests that carbonaceous chondrites may in fact be derived from patches of hydrated and carbonaceous material in normal chondrite material, the host being relatively depleted in the volatile elements which have been concentrated or preserved in these patches, and not actually derived from modification of such material.

The key to the enigma of the carbonaceous chondrites and other extremely rare meteorites may well be found in their rarity. They may simply represent very sparsely isolated patches in large areas of more familiar meteoritic composition. Holyoke may be revealing to us the parent-body situation of carbonaceous chondrites, no more or less. Such occurrences simply cannot be attributed to any process of accretion due to collision in space after parent-body break-up; they must exemplify a heterogeneity pattern of the parent body itself.

We can envisage such patches as the locus of residual carbon and volatile concentration in a small asteroidal parent body; this parent body may well have had less overall volatile content than our own Earth, or lost its volatiles rapidly by degassing violently at an early stage. Just as the Moon, much smaller than the Earth, is now clearly volatile impoverished (Apollo 11 and 12 evidence), so the even smaller parent body to the meteorites may have been even more impoverished. The carbonaceous chondrites are *not*, the author believes, survivals of the fundamental material of the parent body silicate zone, but rather survivals of a paucity of carbon and volatile rich patches within the chondritic material, impoverished by degassing: they were the parent body's apology for the Earth's hydrosphere and atmosphere, a few trapped residues left behind by degassing in the chondritic material of the zone equivalent to the Earth's mantle—material on which they exerted a secondary modification of veining, etc.

15: The Nature and Origin of Chondrules

The chondritic fabric of most stony meteorites has long perplexed scientists. The structure is not really similar to the familiar spheroidal structure of terrestrial volcanic rocks, developed in viscous melts, for only the peculiar fan-shaped chondrules resemble this structure and are characteristically exocentric, the focus of the radial growth being situated outside the chondrule: in contrast, the focal point of radiating growth in volcanic spheroids characteristically lies within the spheroid.

Chondrules may be classified as:

Monosomatic: composed of a single crystal grain.
Polysomatic: composed of more than one crystal grain.

Many of the complex barred forms of chondrules are essentially monosomatic, for the whole chondrule shows simultaneous extinction between crossed nicols when viewed in this section under the polarising microscope. Polysomatic chondrules are more common than monosomatic. Polysomatic chondrules may consist of numerous grains of a single mineral such as olivine or bronzite, or of more than one mineral species, in which case they are mineralogically composite. There is a wide range of textural types, of which the following are the best known:

porphyritic
exocentric fan
barred
grated.

Individual chondrules may not only be mineralogically composite but may show combinations of more than one of these textures—that is they may also be texturally composite. Some chondrules display interstitial glass, which may be clear and translucent, but is com-

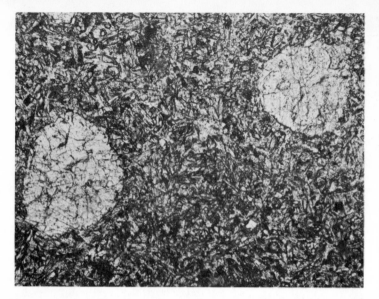

Plate 58 *Chondrule-like spheres of augite glomerocrysts, showing incipient modification of the spheroid to augite crystal outlines, set in a finely crystalline augite-plagioclase-magnetite base. From a basaltic dolerite dyke, Mt Belches, Western Australia. This structure is possibly analogous to chondrule growth*

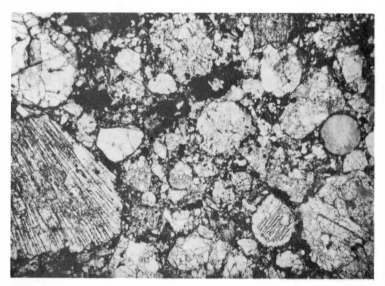

Plate 59 *The Dalgety Downs, Western Australia, olivine-hypersthene chondrite: barred and exocentric fan chondrules are evident in this photomicrograph; the barred chondrule, lower right, is a monosomatic chondrule. The matrix consists of broken chondrules, crystal fragments, and ferruginous amorphous material (after glass?). The grey silicate grains are all olivine or orthopyroxene*

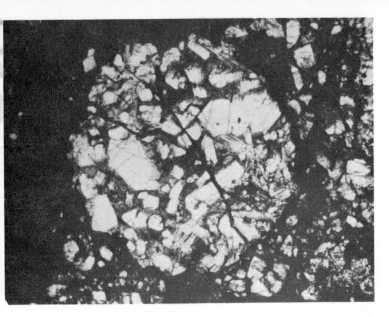

Plate 60 Porphyritic chondrule, with ideomorphic olivine set in glass (partly devitrified) and enclosed in an iron-stained matrix. Frenchman Bay meteorite (C Br), Western Australia

Plate 61 Barred polysomatic olivine-orthopyroxene chondrule; Dalgety Downs meteorite (C Hy), Western Australia

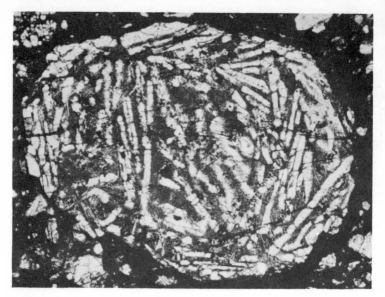

Plate 62 Chondrule showing swirling internal flow texture ('trachytic'); olivine laths in devitrified glass: this could only be derived from a cooling melt. From the Mulga South, meteorite (C Br), Western Australia

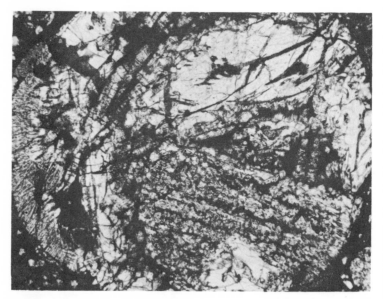

Plate 63 Heterogeneity of texture and composition in a single chondrule. Three textures are represented—porphyritic (upper right), barred (lower right), and radial fan (left margin); the barred area seems to form a chondrule within a chondrule. Three minerals, olivine (white), bronzite (grey), and magnetite (black) are unevenly distributed in the chondrule. Frenchman Bay meteorite (C Br), Western Australia

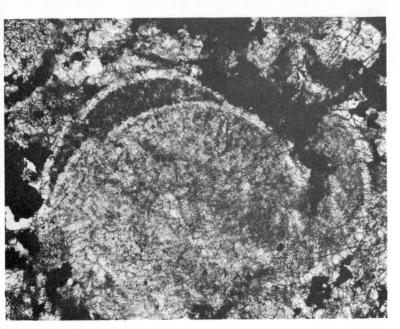

Plate 64 Overlapping chondrules in the Wiluna meteorite, Western Australia (C Br)

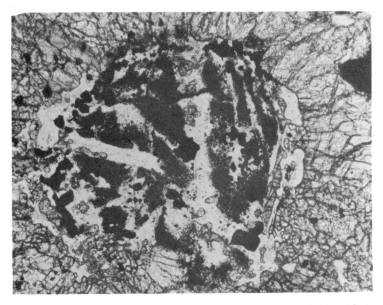

Plate 65 Chondrule containing plagioclase (white laths), troilite, nickel-iron (dark), and devitrified glass (turbid) from the Woolgorong meteorite (C Hy), Western Australia

monly turbid, suggesting devitrification. Chondrules predominantly composed of metal and other opaque constituents, and chondrules predominantly composed of oligoclase feldspar, are known. Chondrules of an essentially cryptocrystalline, amorphous, or glassy character are common, and these tend to have a more glassy core and cryptocrystalline peripheral zone.

Textures displayed by chondrules include the astonishing overlapping structures, and patterns of concentric, trachytic alignment of the olivine laths, a swirling of crystal grains parallel to the border of the chondrule which suggests that solidification in droplet form was the origin. Moulding relationships to other chondrules have also been noted, suggesting a plastic character at the time of formation.

Structures somewhat similar to chondrules have been reported from time to time from basaltic rocks, and basaltic dolerites forming dykes of terrestrial origin. Such structures are essentially spherical aggregations of crystal grains, variants of the more usual glomeroporphyritic clusterings of large inset crystals within a fine crystalline, part glassy, or wholly glassy base material. Though no certain link between such terrestrial growth and meteoritic chondrules has been established, and chondrules may still be justifiably regarded as the 'hall-mark' of meteoritic lithic material, the examples from Western Australia illustrated (see p 192) deserve study. They reflect, in all probability, rapid cooling of a hot magma in a volatile-rich environment, a possible analogy to meteoritic crystallisation. It is interesting to note that the condition of spherical growth in crystal clusters (in effect, a chondritic condition) passes into one in which the outlines of the sphere become modified to those of single augite crystals, despite the composite nature of the sub-spherical growth of augite. It would be interesting to search for examples of meteoritic chondrules transitional into single multigranular phenocrysts in this way.

Theories of Origin

The following theories have been advanced to account for the phenomenon of chondrules:

1 Fused drops of fiery rain (Sorby).
2 Fragments of pre-existing meteorites rounded by oscillation and attrition (Tschermak).
3 Magmatic segregation during the course of rapidly arrested crystallisation of a melt (Brezina).

4 Dispersal of silicate drops from a melt into a hot atmosphere (Wahl).

5 Metamorphism of garnets to enstatite (Fermor).

6 Cooling of molten silicate droplets within a molten rain produced by collision between small and large asteroids (Urey and Craig).

7 Separation in place from a silicate melt (Roy).

8 Liquid condensation in the pre-planetary solar atmosphere (Wood).

9 Spherulitic crystallisation at a comparatively low temperature (1,000 °C) within a homogeneous magma melted under high partial pressures of water and carbon dioxide (Ringwood).

10 Thermal metamorphism of carbonaceous chondrites involving solid state crystallisation at moderate temperatures, with serpentine to olivine transformation (Mason).

11 Direct condensation from a cool dust cloud in which particles were in a colloidal state: the material is supposed to have been originally amorphous and crystallisation only to have occurred after aggregation of chondrules in small planetary bodies (Levin and Sloninsky).

12 Coalescence of spray globules (micro-chondrules) in an incandescent fog, surrounded by a cooler dust envelope, while the chondrules gravitated inwards into the high-temperature zones (Mueller).

Prior's Rules

The applicability of Prior's Rules to chondrites and achondrites and the involvement in these rules of nickel-iron, which besides forming iron meteorites is an almost constant component of chondrites of all types, seems to place severe limitations on the possible genesis of chondrites. If fractionation of the nickel-iron is a process spread over the whole range of meteoritic types then it must be considered in all its implications alongside the separation of the chondrites into a range of classificatory types governed by these rules. Studies by Uhlig and Lovering on the fractionation of nickel-iron in meteoritic irons suggest that the whole range of iron alloys could stem from an initial parental melt of 11 per cent nickel content. Yet complications are added to this rather simple model if one considers not only the irons, but the metal fractions in the stony-irons and stones as well—and surely it is unrealistic not to do so.

Critical meteorite associations are the polymict breccias such as the

Bencubbin stony-iron and the Cumberland Falls enstatite achondrite, both of which contain patches of chondritic material within achondritic host material. Both the sparse chondrite enclaves and the achondrite metal host of the Bencubbin meteorite have been shown by the writer to obey Prior's Rules. This fact alone suggests that the chondrules must represent some form of crystallisation from a melt, and not a solid state process. The texture of the chondrule from the Mulga South meteorite illustrated on p 194 certainly seems quite incompatible with anything but crystallisation from a melt.

Ringwood's idea of spherulitic crystallisation under high volatile partial pressures seems the most compatible with the Bencubbin and Cumberland Falls evidence, which requires the material from which chondrules crystallised to have been differentiated so as to comply with these rules, including the rules governing the partition of nickel-iron ratios in the metal phase, prior to their crystallisation during an early stage in the solidification of the parent body after the post-accretionary heating phase. Only such a history could surely account for the incorporation of chondritic material in the form of rare, small contrasting enclaves, within achondrite/metal material that has surely crystallised from a melt of immiscible fractions within the deep interior of the parent body, and the fact that the enclaves obey the same geochemical rules as the host material. Both Ringwood's concept of degassing on a grand scale early in the history of the parent body or bodies and Mueller's theory of 'microchondrules' (see below) do seem to bear some relation to the observed facts concerning:

1 Polymict breccia associations.
2 Geochemical rules governing such associations and all meteoritic iron/silicate assemblages.

Of the other suggestions, most require no further consideration. The two-stage process, involving prior condensation as chondrules and a subsequent high-temperature stage of recrystallisation, has two forms: one favoured by Wood, in which the chondrules seem to be afforded the status of *planetesimals*, of independent existence prior to accretion of the meteorite parent body from the dust and gas of the solar nebula. This does not seem to be reconcilable with the systematic rules governing the geochemistry of all types of meteorites, and the various elements of polymict breccia associations. The distribution that is represented by these rules can only have been effected in the parent-body stage, as argued above.

In addition, the internal fabric of chondrules may show a degree of heterogeneity quite unexpected in planetesimals—one cannot see how such a degree of heterogeneity could be subsequently imposed without the chondrules losing their discrete entity. Though there can be little doubt, considering the typically igneous porphyritic internal textures and swirling trachytic textures familiar to volcanologists that the chondrules have crystallised from a melt, albeit in strange physical conditions, condensation from a primary nebula gas-dust cloud of droplets to form chondrules would surely never yield anything but reasonable uniform droplets—and thermal recrystallisation could never introduce such internal heterogeneity?

One must suspect from the very nature and appearance of chondrules, which are not unlike little spheroids of volcanic rock, that they are products of rapid cooling of a melt not entirely dissimilar to a volcanic magma, under unfamiliar conditions of hydrostatic pressure and the influence of volatiles forming, perhaps, a temporary 'atmospheric' envelope.

Chondrules and Microchondrules

The theory of Mueller is particularly interesting and deserving of serious consideration. He advances a progressive sequence of development of chondrites, covering the carbonaceous chondrites and three non-carbonaceous classes. He believes that the recognition of the high-temperature mineral olivine in type 1 carbonaceous chondrites indicates that these meteorites are very closely connected, like those of types 2 and 3, to other chondritic matter, which, with the exception of the enstatite chondrites, displays olivine as an abundant constituent. He bases his arguments on statistical petrographic studies of chondrules, a surprisingly neglected field of meteoritic research, and one that offers promise of supplying the badly needed limiting evidence restricting geochemical theorising. Studies of ignition loss in a nitrogen atmosphere at 1,000 °C have also contributed to Mueller's research. He divides the chondrites as follows:

Carbonaceous chondrites Type 1.	Ignition loss 24–30 per cent	Olivine dispersed as small globules or microchondrules set in a quite abundant groundmass of hydrated silicates and carbonaceous matter. The

199

		olivine chondrules commonly show exsolution cores of nickel-iron.
Carbonaceous chondrites Type 2.	Ignition loss 12–24 per cent	Loosely compacted aggregates of microchondrules (referred to by Mueller as 'grape-bunch' aggregates) are the predominant component.
Carbonaceous chondrites Type 3. Enstatite chondrites.	Ignition loss 2–12 per cent	Partly coalesced chondrules are predominant, microchondrules of refractory minerals (olivine and nickel-iron) being embedded in a single crystal or interlocking aggregate of lower melting minerals.
Other chondrites.	Ignition loss less than 2 per cent	Fully coalesced chondrules characteristic.

He deduces that the evolution of chondrules, with decrease in volatile material in chondrites, occurred in two stages:

1 Primary condensation of sparry particles into microchondrules (diameters $\simeq 0.01$ mm).
2 Secondary accretion of microchondrules into chondrules (diameters $\simeq 1$ mm).

Mueller draws an analogy between this model and processes occurring in warm clouds, and has noted that histograms of microchondrules resemble those obtained from fog droplets, whereas those of chondrules resemble those of rain droplets. He envisages a parent body of asteroid size in the form of an incandescent gas cloud covered with clouds of cosmic dust, while at the highest temperature stage of its evolution. The material of type 1 carbonaceous chondrites is supposed by him to come from the interface zone between incandescent fog and cooler dust. The chondrite types containing progressively less volatile material are supposed to originate from deeper and deeper zones in the incandescent cloud mantle, the grape-bunch microchondrules gravitating inwards and coalescing with increasing depth through a rising temperature gradient.

Mueller satisfactorily accounts for the preservation of what appears to be the original spray structure of the asteroid-sized parent body in rare chondrites by invoking solidification of this material at an early stage in degassing, the chondrules being suspended in an enveloping cloud. His theory reasonably accounts for the lack of terrestrial chondritic material, by invoking slower cooling on account of the greater planetary mass, which would allow no early solidification of the spray while the gas remained available to envelope the chondrules —solidification would not, in the terrestrial case, commence prior to the greater part of the degassing. With increasing mass, celestial bodies would thus develop magmatic rather than chondritic textures. There are many loose ends left in the formulation of this novel theory, as in the early stages of statement of all theories, but seems to represent a major advance in our understanding of meteorites.

It remains possible that Mueller's microchondrule → chondrule progression does not represent a vertical progression (ie, reflect depth in the accreting parent body), but that it represents physical conditions in patches within a very unsystematised parent body, and that different compositions of the patches in the parent body, co-existing at the same depth level, determined whether microchondrules or chondrules developed. In particular the presence or absence of carbon and water may have been critical. We find unrecrystallised primitive chondrite material in the Bencubbin stony-iron—may not the very compositional nature of these enclaves have inhibited re-crystallisation? And if so, may not the actual nature of a discrete 'capsule' of carbonaceous chondrite material in normal chondrite material have inhibited full chondrule development? Physicists may very well argue that this is impossible, but there is strong evidence that some such control as this was operative, on recrystallisation at least: and if so, why not on chondrule development? The author sees the possibility that Mueller incandescent fog and transitional incandescent fog/cool dust zones represented by normal and Cc1 chondrites in his model, should be replaced by patches of water, volatile and carbon-rich material, introduced into the product of condensation of the incandescent fog, and perhaps locally inhibiting full droplet 'chondrule' growth.

We must not forget the evidence of Nagy (Chapter 14), that the hydrous material is a subsequent introduction. However, as primitive character of chondrules goes along with the presence of carbon and this hydrous material, it must be supposed to be in some way dependent on its presence—thus, either this material was always present or it was introduced very early, prior to the passage from microchondrule to chondrule stage. The easiest way to evade this difficulty is to see the hydrous and carbonaceous matter as already present in patches, a gas phase, at the time of the high-temperature crystallisation of the microchondrules, inhibiting their full development into chondrules, this phase later condensing, and veining and latering the mass in a deuteric fashion.

Summing Up

Mueller's observations have been reconciled with condensation in the early parent body stage, and this, in the author's view, is the earliest possible time for microchondrule and chondrule formation. The author accepts these observations but relates them to a concept of a heterogeneous parent body, patchy in its outer 'fog' envelope (=chondrites), as well as in its inner zone (=achondrites and subordinate metal and rare chondrite enclaves), and in the core (= metal alloys)— a patchiness that he regards as indicated by the nature of meteoritic polymict breccias. Intermingling of material in the melted zones by the agency of a stirring up of diverse melts must have occurred, subsequent to the post-accretion heating and internal melting, and material must also have been aggregated in strange assemblages by disruptive disturbances premonitory to parent body disruption. Due to such processes, the internal matter of the parent body was so jumbled up that chondritic material from the fog condensation zone of Mueller could get incorporated in the still molten alloy/achondrite matter of the stony-iron meteorite (Bencubbin) and into the achondrite material of the Cumberland Falls meteorite. To suppose that such polymict breccia associations, however inconvenient they are to theory, were *not* formed in the parent-body stage, but were derived by later chance collision of meteoritic material, is to ignore the evidence to the contrary.

Experiments (reported on by Nelson, L. S., Blander, M. and Skaggs, S. R., in NASA preprint SC-DR-71 4113) reproduced chondrule-like forms from free-falling droplets of various silicates melted by CO_2 laser. Rapid crystallisation in a supercooled melt is indicated: and either small scale volcanic, impact splattering or electrical discharge events in a low temperature ambient medium: or even slower cooling of droplets during condensation of the nebula of solar composition in a high temperature ambient medium.

The discovery of chondrules of anorthositic norite composition in Apollo 14 lunar samples has led to attributions to impact melting and splattering with spontaneous crystallisation of highly supercooled, freely floating, molten droplets (Kurat, G., Keil, K., Prinz, M., and Nehru, C. E., unpublished paper): but as yet no evidence to discount the alternative possibility, that crystallisation of lunar chondrules from molten droplets occurred during rapid internal degassification involving a primary ambient gas medium.

16: The Metamorphism of Meteorites

The term metamorphism, widely applied to meteorites, is perhaps unfortunate, because the processes embraced by this term in meteoritics are not very similar to the metamorphic processes familiar to geologists—*contact metamorphism*, induced by proximity to a hot igneous intrusion, and *regional metamorphism*, a widespread process combining effects of increased temperature, largely due to depth of burial, and directed pressure related to tectonic stresses.

In the meteoritic context, this term denotes *thermal recrystallisation*, and in fact, the term *recrystallised* is generally applied to thermally modified chondrites.

Three types of thermal effects can be naturally imposed on a meteorite mass:

1 Effects within the parent body, before separation of fragments orbiting in space as individual meteoroids.
2 Effects on meteoroids orbiting in space after separation from the parent body.
3 Effects on meteoroids while heating up on entering the Earth's atmosphere and ablating.

In addition to these natural thermal effects, the possibility of artificial thermal modification of irons during attempts to forge them is a real one: several irons show evidence that this has occurred. The meteorite mass may be modified further by effects of shock on hitting the surface of the Earth, and by terrestrial weathering, but neither of these processes normally effects significant thermal modification.

Thermal modification while still within the parent body seems to be by far the most important of these thermal modification processes, and the greater part of the recrystallisation of meteorites may be attributed to such a process.

The types 1 and 2 carbonaceous chondrites show little evidence of

205

recrystallisation, though the carbonaceous chondrites appear to have undergone more than one episode of superimposed low-temperature modification involving hydration. The non-terrestrial water content of carbonaceous chondrites displays a peculiar isotopic pattern: it can be driven off completely at $180\,°C$, and this imposes a limit to the temperature at which they were formed, or at least attained the form familiar to us. Studies of the strained glass fraction of the Mighei carbonaceous chondrite showed that it could not have been subjected to temperatures above this level for a period of more than two weeks, or above $250°$ for an hour, and it could not have been raised above $300°$ even momentarily. Despite this evidence, the metal fraction in nine carbonaceous chondrites was found by Wood to be compatible with equilibration at temperatures as high as $1,400°$ to $1,535°C$. One solution to this anomaly is to suppose that the metal solidified prior to its aggregation with the low-temperature components of the meteorite (or its invasion by them) to form a mechanical aggregate of high- and low-temperature mineral phases, and that there was no subsequent heating. The metal fraction apparently cooled very rapidly, for Wood notes that kamacite and taenite have not separated out from one another.

An alternative explanation is that the volatile elements in carbonaceous chondrites were enscapulated—that these chondrites, as suggested previously, represent patches of carbon, water, and volatile-rich material in the normal chondrite envelope of the parent body: that is, patches in which the volatiles were retained in the gas phase during high-temperature crystallisation of silicates and metal *because they could not escape*. The volatiles would only have been driven out at high temperature from a 'condensing incandescent fog' if they could find escape, and perhaps the denser surrounding 'incandescent fog', or liquid or solid, inhibited this escape.

Most of the common chondrites show varying degrees of thermal recrystallisation, involving much higher temperatures than the carbonaceous chondrites could apparently have suffered. Dodd, van Schmus, and Koffman have erected a small primitive group of common chondrites, referred to as the *unequilibrated common chondrites*. These show nothing but the barest traces of thermal metamorphism, and a variation in the composition of mineral phases that are members of isomorphous series (olivine and pyroxene), indicating lack of attainment of chemical equilibrium. Both olivine-bronzite and olivine-hypersthene chondrites are represented in this small group,

206

and it also includes members of the H, L, and LL total iron types of Urey and Craig and Keil and Frederiksson.

The unequilibriated common chondrites include:

LL Ngawi. Semarkona.
L   Krymka. Bishunpur. Hallingeberg.
H   Sharps. Tieschitz.

The carbonaceous chondrites of type 3 (also known as the olivine-pigeonite chondrites) all show appreciable evidence of thermal metamorphism, with the exception of the Vigarano meteorite. Some enstatite chondrites show little trace of thermal metamorphism, whereas others show appreciable metamorphism.

Van Schmus and Wood erected a classification of meteorites according to metamorphic types in 1967. This classification was intended to be used alongside the usual Prior–Mason classification, not to replace it. It has six divisions that are supposed to represent degrees of progressively increasing metamorphism. There has been

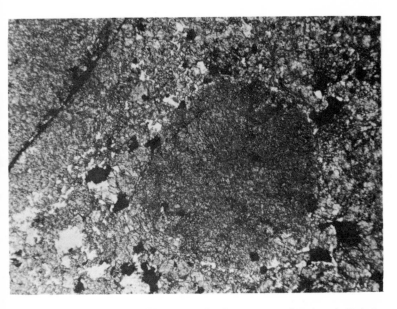

Plate 66 Chondrule in a highly recrystallised meteorite—oligoclase (white) is unusually abundant, as is not uncommon in such recrystallised meteorites, Woolgorong meteorite (C Hy), Western Australia

TABLE 18

Chondrites—Additional Mineralogical and Petrographic-textural Classification of Van Schmus and Wood, 1967

	Petrologic types					
	1	2	3	4	5	6
(i) Homogeneity of olivine and pyroxene compositions	—	Greater than 5% mean deviations	Greater than 5% mean	Less than 5% mean deviations to uniform	Uniform	
(ii) Structural state of low Ca pyroxene	—	Predominately monoclinic	Predominately monoclinic	Abundant monoclinic crystals	Orthorhombic	
(iii) Degree of development of secondary feldspar	—	Absent		Predominantly as microcrystalline aggregates		Clear, interstitial grains
(iv) Igneous glass	—	Clear and isotropic primary glass; variable abundance		Turbid if present	Absent	
(v) Metallic minerals (maximum Ni content)	—	(<20%) Taenite absent or very minor	Kamacite and taenite present (> 20%)			
(vi) Sulphide minerals (average Ni content)	—	>0·5%			<0·5%	
(vii) Overall texture	No chondrules	Very sharply defined chondrules		Well-defined chondrules	Chondrules readily delineated	Poorly defined chondrules
(viii) Texture of matrix	All fine-grained, opaque	Much opaque matrix	Opaque matrix	Transparent microcrystalline matrix	Recrystallised matrix	
(ix) Bulk carbon content		0·6-2·8%	0·2-1·0%		<0·2%	
(x) Bulk water content	~20%	4-18%		<2%		

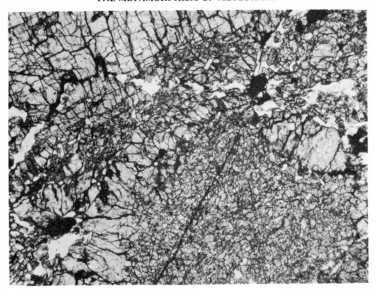

Plate 67 Chondrule outline barely discernible: very highly recrystallised meteorite, Yayjinna meteorite (C Hy), Western Australia

consideraole doubt whether the classes 1 and 2 of van Schmus and Wood really represent the same progressive sequence of metamorphic changes as the other four classes. These two classes refer to the carbonaceous chondrites of types 1 and 2 only. Class 3 covers the unequilibriated chondrites and those enstatite chondrites that have suffered minimal thermal metamorphism, and classes 4 to 6 cover progressive stages in the modification of the thermally recrystallised common chondrites and enstatite chondrites.

The classification is given in Table 18.

This additional classification certainly makes the task of the petrographer of chondrites much more meaningful: hitherto, petrographers had been aware of these changes, but the sequential relationships were not defined. It must be noted, however, that some of the criteria require refined and expensive instrumental techniques—for example the determination of the homogeneity of the mineral phases requires the use of an X-ray diffractometer or microprobe—not equipment available to all petrographers! In addition, there is some doubt about the validity of probe results of this kind at the

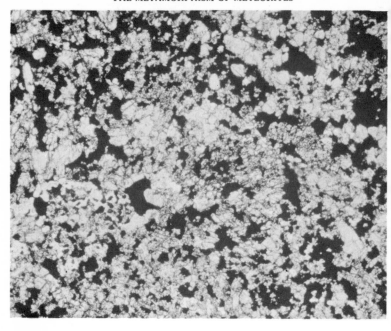

Plate 68 There is a chondrule just discernible in the centre of the photograph, in this almost totally recrystallised meteorite, West Reid meteorite (C Br), Western Australia

present time, and though variation may be well established in individual mineral phases, the actual range of values may be little more than an approximation because of uncertainties involving standards.

Among the progressive effects that are easily determined are progressive loss of chondrule outlines, which in most cases are still discernible, if obscure, even in highly recrystallised chondrites. Mason, however, has described the Shaw, Colorado, chondrite, a highly recrystallised olivine-hypersthene chondrite in which no trace of chondrules is preserved. Another such effect is the loss of the lamellated clinohypersthene mineral phase, it being progressively replaced with increasing thermal recrystallisation by orthopyroxene and plagioclase. Thus the more highly recrystallised chondrites contain plagioclase in the form of clear, interstitial grains of very small dimensions, grains which may or may not show lamellar twinning (and may be converted to maskelynite).

Van Schmus and Wood do not suppose that a type 6 chondrite in

210

their classification *necessarily* represents modified type 5 material, and so on. Their table would be more logical if types 1 and 2 were omitted, for they do not seem to represent progressive stages in the same metamorphic sequence like the other types. However, the table, as it stands, does cover all chondrite textures and is quite acceptable as long as it is realised that types 1 and 2 represent carbonaceous chondrites, and that types 3 to 6 are applicable to other chondrites, and C_3, C_4 carbonaceous chondrites.

Achondrites do not show equivalent thermal recrystallisation patterns, yet they are clearly not simply recrystallisation products derived from chondrites, despite the similarity between certain achondrite types and chondrite types in their bulk chemistry and mineralogy (for example, diogenites and olivine-hypersthene chondrites: aubrites and enstatite chondrites): there are too many points of disparity in their chemistry for this to be the case. The Shergottite category of achondrites is simply a shock conversion of a eucrite, the ophitic texture being preserved but the feldspar being entirely converted to the isotropic mineraloid *maskelynite*. Similarly the granulated eucrite material present in the Mt Padbury mesosiderite as enclaves, and also seen in a new eucrite from Millbillillie, Western Australia, represents a shock granulation (in the parent body)* without appreciable thermal modification. These effects are neither ablation nor impact shock effects, nor are they thermal recrystallisation effects.

A critical fact that needs explaining is why there is a complete absence of transitional types between achondrites and chondrites. That chondrites do not simply pass with increasing recrystallisation into achondrites seems certain, considering the evidence. There seem to be three quite distinct processes evident in stony meteorites:

1 Microchondrules—chondrules (seen in Cc 1 and Cc 2 mainly).
2 Chondrules—obliterated by thermal recrystallisation (seen in CEn, CBr, CHy, Cc 3).
3 Crystallisation of achondrites, direct from silicate melts.

What seems to be evident is that none of these three processes is the extension of another. The meteorite parent body seems to have been divided into zones of quite different crystallisation behaviour, and by discontinuities across which there was little or no movement and little influence exerted. Whether these zones were of the scale of

* The enclosing metal may be unshocked in Mt Padbury.

concentric shells or simply discrete irregular patches, within a very crude shell configuration, remains debatable.

The iron meteorites may also display recrystallisation. Thermal alteration has modified the structure of hexahedrites producing the so-called nickel-poor ataxites in which the Neumann lines have been obliterated: similarily there is a group of fourteen *metabolites*, meteorites of octahedrite composition in which the Widmanstätten pattern has been thermally obliterated.

It was at one time popular to explain such alteration of irons by invoking thermal heating on close approach of the meteoroids at perihelion to the Sun's heat. There seems to be no evidence, however, that this effect is not due to earlier quite prolonged thermal modification occurring within the parent body, anologous to the recrystallisation process affecting chondrites. The evidence for appreciable heating of meteoroid masses while orbiting as individual entities in space seems to lie purely in interpretations of geochronological evidence based on radioactive mineral decay studies (Chapter 18). There is no certainty that the inferences drawn are correct, and there is thus no certainty that significant heating up and thermal alteration during the meteorite orbiting phase is a reality, though it may well be a complicating thermal modification factor.

Ablation heating during entry into the Earth's atmosphere produces only a melt skin, a millimetre or so thick, and possibly penetration of fused material into cracks in the mass. The thin dermal layer of stones may form three distinct tiers. In rare cases apparent regeneration of olivine has been noted in fusion crusts, for example in the Wiluna chondrite which fell in Western Australia in 1967.

Irons also show dermal ablation zones in which their textures are modified: hexahedrites show dermal bending and obliteration of Neumann lines, whereas octahedrites show obliteration and deformation of their Widmanstätten patterns and the ataxites similarly show a structureless zone, free even of the minute kamacite crystals.

17: Organic Matter in Meteorites and Possible Primitive Life Forms

'Proof that organic compounds and organised elements in meteorites are the residues of living organisms indigenous to the meteorites would be the most interesting and indeed astounding fact of all scientific study in recent years'—H. C. Urey.

The story of the discovery of organic matter in meteorites starts in 1834, with Berzelius, a Swedish chemist, and the fall of the Alais carbonaceous chondrite in Southern France. Berzelius discounted the possibility that the carbonaceous material which he discovered represented extra-terrestrial life forms. Though the sooty residues in carbonaceous chondrites, a largely non-volatile and non-soluble mixture of organic compounds of considerable complexity, were studied at intervals throughout the century that followed, and another meteorite of similar type carrying these mysterious compounds fell at Orgueil, France, in 1864, it was not until 1961 that there was a sudden explosion of interest in this aspect of meteoritics.

Nagy, Hennessy, and Meinschein, in 1961, reported that the volatile fraction of the organic material consisted largely of saturated hydrocarbons characteristic of ancient terrestrial sediments and petroleum. The solid particles and distillates included paraffins and cyclic hydrocarbons. The relative abundances of compounds of different molecular weights in each group were compared to the abundances in butter and recent terrestrial sediments carrying organic material of biogenic origin. A very close match was obtained.

It is quite clear that the hydrocarbons concerned are of extra-terrestrial origin—their quantity and consistent recognition in carbonaceous meteorites is such that they must have come in with the meteorite. What is not at all certain is that they are of biogenic origin. In fact the most recent research on the subject has tended to support the view of Anders and his associates, that they are of extra-terrestrial origin, but not of biogenic origin, though they may be the

213

primitive material from which life, by some completely obscure trick of nature, was initially synthesised in its most simple, primitive forms.

The survival of this material in spite of ablation heating presents no problem—Mason has drawn a comparison to a dish known as a 'baked alaska', and it is quite certain that ablation only affects the periphery even of quite small masses like the carbonaceous chondrites (maximum type 3, Vigarano, Italy, 1910, 16 kg (up to the fall in 1969 of the $\geq$ 1 ton Allende recovery): maximum type 2, Murray, USA, 1950, 12 kg: maximum type 1, Orgueil, France, 1864, 11 kg: smallest, Tonk, India, 1911, 0·01 kg).

The first really modern analytical work on the Cold Bokkeveld carbonaceous chondrite of type 2, by Mueller, in 1953, was the precursor of the more detailed spectrographic investigation by Nagy *et al*. Mueller recognised 19·84 per cent carbon, 6·64 per cent hydrogen, 3·18 per cent nitrogen, 7·18 per cent sulphur, 4·81 per cent chlorine, 40·02 per cent oxygen and associated elements, and 18·33 per cent ash in a 1·1 per cent residue of resinous material from the sample, after free sulphur separation from it. The nitrogen, chlorine, and sulphur are characteristic components of complex organic compounds, and Mueller concluded that the nature of the resinous residue was a mixture of these compounds with free sulphur.

Organic compounds are not necessarily of biogenic origin, in spite of their name. Many have been artificially synthesised from inorganic compounds. The identification of aromatic compounds very similar to those in organic material in recent sediments, by Meinschein, provided strong, but by no means conclusive, evidence of biogenic origin.

The case was complicated in 1961 by the report by Claus and Nagy of the discovery of microscopic particles resembling fossil algae in the Orgueil and Ivuna meteorites. This sparked off the renowned 'museum contamination' controversy. While it is quite clear that there *are* many terrestrial contaminants in specimens held for years in museums—by local bacteria, spores, and pollen, and exotic contaminants brought in with other specimens from other parts of the world—and the porous nature of the carbonaceous chondrites makes them ideal storage receptacles for such contaminants, it is also quite certain that there *are* microscopic bodies of apparently organised structure, which *are* a primary feature of the meteorites. But are they life forms? Inorganic processes can produce bodies of astonish-

214

ingly organised appearance—we have only to consider the complexity of snow-flake crystals, and, in fact, one authority has called these bodies in the meteorites 'carbonaceous snow-flakes'!

Tests with staining agents, suitable for detecting DNA (deoxyribonucleic acid) were at one time supposed to provide positive evidence for biogenic origin, but, alas, the same results were obtained on the igneous rock kimberlite! Tests for optical polarisation, a typical feature of organic compounds of biogenic origin, were applied, first by Mueller to his residues, and were negative; however, since then tests carried out in completely sterile laboratories by very sensitive methods have shown optical polarisation, but of a type quite unlike that of terrestrial organic compounds of biogenic origin.

Taking the Nagy and Anders examination of the problem up to the present time it is profitable to conclude with an extraction of the main points of statements by Nagy (1968) and Hayes, an associate of Anders (1967). Nagy notes the evidence for a hydrous, low-temperature environment for the type 1 carbonaceous chondrites—it is generally agreed, he states, that you cannot synthesise hydrated silicates without water in either liquid or vapour phase. There is in fact evidence of repeated leaching by aqueous solutions. He concludes that these environments are not at present understood. He expresses some support for the extra-terrestrial origin of the optically active 'lipid' components. Though admitting that there is a possibility that this activity may be due to instrumental artifacts, and that these compounds are similar in their $^{13}C/^{12}C$ isotopic ratio to Miocene and Pennsylvanian crude petroleums of marine origin and the lipid fraction in marine plants, he also notes that such material is an unlikely contaminant of dry museum shelves! Even if some uncommon non-marine micro-organism does contain lipids with such an isotopic ratio, a problem remains in the lack of amino acids which would be expected if these organisms died in situ after the food supply ran out. He considers that the nature of the organic substances is compatible with a biogenic or a non-biogenic origin, in spite of the resemblance of some components to biogenic compounds.

Using scanning electron microscopy Nagy has shown that the 'organised elements' may be partially embedded in fresh broken rock surfaces, or may have mineral particles embedded in them. They are clearly not terrestrial contaminants. Nagy notes the three fiercely argued explanations—(a) abiological, (b) extra-terrestrial, biological, (c) terrestrial, biological. He preserves a quite admirable open mind, quoting Urey, 'if found in terrestrial objects, some substances in meteorites would be regarded as indisputably of biogenic origin' and expressing his view that there is strong evidence in favour of an extra-terrestrial biogenic origin. He concludes by hoping that the lunar samples, to be brought back in the near future (and now actually brought back by Apollo 11, 12, 14, and 15), will not engender, on analysis, such long lasting debates! These lunar samples have, in fact, yielded no such controversial material (Chapter 23).

Considering the organic compounds, Hayes notes that the optical evidence is inconclusive. Amino acids are probably absent. Traces of nucleic acid bases reported by Hyatsu in 1964 were accompanied by compounds of no biological significance (melamine and ammeline). The porphyrins detected by Hodgson and Baker in 1964 are only in very minute quantities. The correlation with terrestrial sedimentary hydrocarbons is, however, considered well established, as is the recognition of pristane and phytane, isoprenoid hydrocarbons (Meinschein, 1964). These are the only two compounds in a lengthy list of those recognised, that have not been abiogenically synthesised. Further, lack of attempts to synthesise such compounds means that their presence is far from strong evidence of biogenic origin.

The widespread occurrence of carbon in meteorites points to an origin in a primary process during the formation of the solar system, and the quantitative correlation of carbon content with certain trace element concentrations supports this interpretation. The most promising theories for meteorite genesis introduce models involving asteroidal parents, and there is no invocation of an extra-terrestrial biosphere or atmosphere in these models. The most weighty evidence against biogenic origin for the organic compounds, if one accepts one of these models, is the fact that the nature of the processes involved seems quite incompatible with the existence of life forms. Taking the now scarcely tenable lunar provenance models once favoured by Urey, and accepting some transfer of a primitive terrestrial biosphere at the time of capture or ejection of the Moon, a biogenic origin seems remotely possible, but these models are far-fetched (Chapter 23), and scarcely worth consideration.

Similarities between terrestrial and meteoritic hydrocarbons noted by Oro et al. in 1966 could favour such a model—or even a model for terrestrial provenance of some meteorites! However, Hayes notes that the carbonaceous chondrites cannot simply be derived by addition of water, organic compounds, and trace elements (in cosmic proportions) to ordinary chondrites, and that the differing distributions of the various organic components in carbonaceous chondrites of the three types also has to be explained. He poses a number of questions in conclusion: Are the variations in certain organic compound (n-alkanes) distribution between the three types related to temperature conditions or retentivity? Is there a relationship between low n-alkane content and higher n-alkane/isoprenoid ratio in two type 2 carbonaceous chondrites (Renazzo and Kaba) and an increase in metal in their matrices? Is there any relationship between organic content and degree of metamorphism (thermal recrystallisation)?

Do H and L types of unequilibrated common chondrites show differences in organic contents? Do ordinary chondrites contain organic material—or is the evidence of polymeric material in Holbrook (noted by Hayes and Birmann in 1967) due to extremely fast cooling? Do reheated chondrites contain a unique type of distribution of organic material or compounds? Hayes was suggesting, in posing these questions, that there are indications that we should be moving from a search for extraterrestrial life into a study of organic cosmochemistry, for the evidence suggests that these organic compounds reflect cosmochemical rather than biogenic processes.

The recent studies by Ponnaperuma and others at Ames Research Centre on the Murchison Cc2 meteorite that fell in Victoria on 28 September 1969 confirm this view. Neither amino acids nor other carbon compounds found in this meteorite appear to be of biological origin. The group concluded that such acids can form and evolve by purely chemical means in Nature, outside the Earth. The meteorite was reportedly found to be 4,500,000,000 years old (Anon. *Geotimes*, **16**, (1), 14, 1971), and these compounds were thus apparently available at this early cosmic date; and similar materials may have contributed to the origins of life in the solar system, at a far earlier date than geologists commonly accept. Five of the twenty acids were those found normally in living cells: however, another eleven such acids, normally present in living cells, were not identified. These acids, in the Murchison meteorite, were almost equally of right- and left-handed structure, whereas amino acids produced by terrestrial biological organisms are only very rarely right handed.

The presence of organic material is not limited to carbonaceous chondrites—an enstatite chondrite, ureilites, unequilibrated common chondrites (H and L types), and the troilite nodules in irons have all revealed such compounds. The high-temperature stage through which the irons must have passed means that in the latter case some reaction between gaseous components within the nodule must be invoked, if these are indigenous and of extra-terrestrial origin. There are similarities between these compounds and those of carbonaceous chondrites which give support for the cosmochemical origin favoured by Hayes. Some years ago a somewhat original study was made by Sisler, who actually cultured samples of the Murray carbonaceous chondrite in a germ-free laboratory and obtained growth in his culture media after several months! It seems highly likely that this was not evidence of viable micro-organisms, but that Sisler was actually culturing museum contaminants! In spite of the attractively organised form of some structures in carbonaceous chondrites, and the fascinating by-ways of research opened up by them, the biogenic concept may well be in the nature of a 'red herring' as Hayes implies; even if there is no certainty of this and the proof mentioned by Urey in the quotation that heads this chapter may yet be forthcoming! One thing is certain though—it is not with us yet, and if, anything, the tide is running out from the sensational biogenic theory towards the less sensational, but scientifically more profitable, shallow waters of cosmochemical theory.

18: Meteorite Ages and Isotopic Studies

According to Anders, on whose review of the subject this chapter is based, five events can be dated by radioactivity:

	Method used	Isotopes used
Fall (arrival on Earth)	Cosmic-ray induced radioactivity	^{39}Ar, ^{14}C, ^{34}Cl
Break-up of parent bodies		^{3}H, ^{3}He, ^{34}Cl, ^{34}Ar, ^{38}Ar, ^{39}Ar, ^{40}K, ^{41}K
Cooling of parent bodies *Melting of parent bodies*	Long-lived radioactivity	^{238}U, ^{40}K, ^{87}Rb, ^{187}Re, etc
Nucleosynthesis	Extinct radioactivity	

Fall

The date of fall can of course be theoretically dated to some extent by measurements of the age of the terrestrial geologic formation in which a meteorite is enclosed when found: however, besides uncertainties due to possible reworking, the fact is that, unlike tektites, meteorites are very seldom found incorporated within datable geologic formations. Their recovery is limited to surface finds or shallow burials in soils of recent age. A recovery of a meteorite from a borehole core in America is reported, but authenticated cases of containment in older formations are, surprisingly, virtually absent from the record. Thus, the only real possibility of dating the arrival on Earth of finds is by measuring the state of progress of unsupported decay of isotopes produced by cosmic-ray exposure, between the time of parent body fragmentation and Earth arrival. These results indicate

219

that all known meteorites stem from falls that took place no earlier than a million years ago. The record of meteorites in the form of available specimens thus comes from virtually the last instant of the 3,500 million year time span recorded by terrestrial rocks. There is no certainty that meteorites arrived on Earth throughout this longer time span: the present flux could well be an unusual event in cosmic history.

Meteoroids in Space were irradiated by cosmic rays up to their arrival on Earth. Such rays affected different parts of the mass to varying degrees, for the outer layers shield the interior from them. Before disruption of the parent body, only its surface would be subject to cosmic ray exposure, the interior being shielded. These methods are therefore theoretically only applicable to material from the interior of the parent body. We assume that meteoritic material is just this, in applying the method. It is assumed that one can extrapolate the present intensity and spectrum of cosmic rays backwards in time and assume a constancy for the cosmic ray flux in Space throughout the solar system. Thus these methods rely on certain assumptions, some solidly founded, some necessary but not justifiable with any degree of certainty. Results obtained must be treated with reserve because of these assumptions, necessary to render the methods practicable. There are, however, certain patterns in the actual results obtained that can be used to test the validity of some of the assumptions, so these methods are not quite so dubious as might at first appear.

An iron nucleus may, for example, be struck by a high-energy proton. It loses several particles of light mass (protons, neutrons, helium nuclei) in the process. The result is to leave behind stable and radioactive residual nuclei, which are subject to measurement. The quantity of these nuclei present can yield information concerning the date of fall of a meteorite, and also concerning its history between parent body break-up and fall to Earth. After the fall the meteorite is shielded from cosmic rays by the Earth's atmosphere, and so, from arrival on Earth onwards, the cosmic-ray induced radionuclides will decay in an unsupported manner according to their particular half-life (the time taken for decay to reduce the amount of the parent element to half its initial value).

The radionuclides:

^{39}Ar   half-life 325 years

^{14}C half-life 5,760 years
^{36}Cl half-life $3\cdot08 \times 10^5$ years

are the only ones that have been used to date meteorite falls.

For theoretical reasons (because of the number of unknowns in the equation applicable) two such radionuclides are best used together. ^{39}Ar and ^{36}Cl are so used. However, the radionuclide ^{14}C can be used alone, an assumption for the value of one of the unknown quantities being made on the basis of a check between calculated results and known dates of falls of a number of meteorites, the value so obtained being then used for meteorites of unknown fall data.

The results of applying this method suggest that irons weather very slowly. The Washington County, Colorado, iron has a calculated fall date of 1,400 years ago, but is described in the literature as very fresh, a disc-shaped mass, possibly fell about 1916! Another iron from Keen Mountain, Virginia, calculated to have fallen 1,290 years ago, has a very fresh appearance and was supposed to have fallen in 1940! It is apparent that some irons may be very old indeed. The Williamstown, Kentucky, and Grant, New Mexico, irons according to calculation fell nearly a million years ago. Stones yield very low values, virtually zero. This is in accord with the known rapidity of weathering of stones, but the supposition that they weather away in all conditions between 1 and 100 years after fall is not borne out: chondrites from Potter, Nebraska, and Woodward, Oklahoma, have calculated ages of fall as high as 20,000 and 14,000 years ago respectively.

The method has limited application in correlating recent finds with known falls in the same area, long ago, and might be used to geologically date meteorite craters—for example by application to the Wolf Creek, Henbury, Boxhole, and Cañon Diablo irons.

Cosmic-ray Exposure Ages

Cosmic-ray exposure only commenced in the case of meteoroids when they became distinct identities, metre-sized or smaller fragments separated from the parent body: prior to this they were shielded within the parent body. From a study of both stable and unstable isotopes resultant on cosmic-ray exposure it is possible to calculate the date of break-up of the parent body. This is known as the *radiation* or *exposure age*. The size and shape of the meteorite between break-up of parent body and fall to Earth is assumed to be constant, that is, the amount of shielding remained constant. This

221

assumption is not always justified—there is evidence that the Odessa, Texas, and Sikhot-Alin, USSR, irons suffered intermediate break-up while orbiting in Space.

The theory requires the use of more than one cosmogenic nuclide and among those used are the isobaric pairs ^{3}H and ^{3}He: ^{36}Cl and ^{36}Ar: and the isotopic pairs ^{39}Ar and ^{38}Ar: ^{40}K and ^{41}K. Similar methods, ^{39}Ar coupled with ^{45}Sc, and ^{39}Ar coupled with ^{3}He, have also been investigated.

The method has also been supposed to be capable of showing the comparative degree of shielding a sample, and, by a method of contouring, indications have been obtained of the form of the mass before atmospheric ablation or disintegration. In certain cases the percentage of ablation loss can be calculated. The results obtained for the exposure ages are, however, confusing, though some degree of system emerges. The nature of this systematisation is unexpected. Irons tend to have much higher exposure ages than stones. Irons show clustering according to two peak values:

$$0.50\text{--}0.59 \times 10^9 \text{ years.}$$
$$0.24\text{--}0.29 \times 10^9 \text{ years.}$$

Irons also show a grouping of ages according to metallurgical or structural class:

$$\text{Medium octahedrites} \geq 0.3 \times 10^9 \text{ years}$$
$$\left.\begin{array}{l}\text{Hexahedrites} \\ \text{Ataxites}\end{array}\right\} \quad \leq 0.3 \times 10^9 \text{ years.}$$

Anders notes that this surprising result is not incompatible with the observation that all hexahedrite falls occurred between midnight and noon, whereas all but one octahedrite fall occurred between noon and midnight—implying some distinction in orbital characteristics corresponding to metallurgical or structural class.

Irons and stones show a systematic disparity:

$$\begin{array}{ll}\text{Stones} & 20\text{--}30 \times 10^6 \text{ years} \\ \text{Irons} & 10^8\text{--}10^9 \text{ years.}\end{array}$$

Paneth assumed a single parent body for all meteorites and deduced a single break-up event, 2×10^9 years ago. If this were true all meteorites should have similar cosmic-ray exposure ages. The results obtained do not really conform to the expectation of the commonly accepted modern theory of multiple parent body break-up events;

which, however, for other reasons still finds favour. The expectation is an age spectrum with a peak value and continuum extending towards lower values. A model involving several primary break-ups followed by numerous collisions of meteoroids is the most favoured one, and Kuiper suggests that there were continuous collisions in the asteroid belt. This is statistically likely, but would lead to a complete dampening out in the fluctuations in production rate due to primary asteroidal collisions, because it would inevitably result in innumerable secondary meteoroid collisions. Against this concept must also be ranged the systematic disparity between exposure ages of stones and irons.

Urey believed that, because average ages closely matched deduced collision life-times for objects coming from the Moon and the asteroidal belt respectively, irons plus a few stones possessed of anomalously high exposure ages came from the asteroidal belt, whereas stones and a few irons with anomalously low exposure ages had been spalled off the lunar surface as a result of impacts (see further discussion, Chapter 13). This and similar lunar provenance hypotheses cannot stand up to rigorous analysis. However, as Anders remarks, no better explanation has been advanced. Firemen, Felice, and Whipple advocated some form of space erosion by interplanetary dust, but the ages calculated for meteorites of high and low microscopic crushing strength do not support this suggestion: and the property of microscopic crushing strength appears to show uniformity throughout the entire meteorite type range: this property is critical as cosmic dust is so fine as to be only able to erode by attacking constituent mineral grains.

Satisfactory explanations for the nature of the results, especially the two clusters for irons and separation according to metallurgical types, and the clustering of chondrite stones at 22 million years, have not yet been advanced.

Anders considers the possibility of sorting of meteoroid types in different positions in Space after break-up of the parent body: for it requires very special circumstances to produce the highly elliptical Mars-passing orbit extending to within one astronomical unit of the Sun at perihelion, an orbit that can bring meteorites to the Earth. The meteorite collections held on Earth may well be a highly selective representation of the whole range of meteoritic material in Space. Space erosion may be a minor factor in modifying exposure ages, but some other explanation for the anomalies must be sought.

223

The Size and Shape of Meteoroids Before Ablation, and Percentage Ablation Losses

The use of cosmic-ray produced isotopes to determine the pre-atmospheric entry size and shape of meteoroids from which meteorites have been derived by ablation and, in some cases, atmospheric fragmentation, has been investigated by a number of workers. It is obvious that ablation losses will depend on geocentric velocity and on the nature of the mass, so there will be a wide variety in the degree of ablation loss between individual meteorites. The method can also be used to contour a mass, and give an indication of its shape in Space.

Degrees of ablation loss calculated by various methods are given in Table 19.

TABLE 19

Ablation Losses After Anders (1963)

Method	Ablation Loss value (%)	Class	Meteorite	
^{3}He—^{4}He	73	Off	Grant, New Mexico	Irons
^{3}He	45			
Metallographic	20–48			
^{3}He—^{4}He	92	Om	Carbo, Mexico	
He—Ne	78			
^{3}He—^{4}He	86	Om	Casas Grandes, Mexico	
He—Ne	78	Og	Mt Ayliff, S Africa	
^{3}He—^{4}He	99	H	Keen Mtn, Virginia	
Metallographic	27	A	Tawallah Valley, Northern Territory, Australia	
Metallographic	60	A	Wedderburn, Victoria, Australia	

Solidification Ages

Extensive melting is believed to have occurred within the parent body some time after its formation by cold accretion from the dust of the solar nebula (the accretionary model for the origin of the solar system, despite objections to it, is accepted, tacitly, as a basis for all meteoritic age-dating studies). The melting of the parent bodies cannot be solely due to radiogenic processes as was at one time supposed, but must involve other processes—for instance, gravitation inwards of the last accreted heavy metallic fraction as suggested by

Ringwood: there may even be quite unsuspected additional post-accretion heat sources such as internal tidal effects. Whatever the source of the heat, the evidence suggests that heating occurred soon after the formation of our own planet and the other planetary bodies, including the asteroids, producing general internal melting with partial separation of metal and silicate phases, together with other chemical fractionations. The accretion stage—the actual formation of the asteroidal parent body—is not datable by any known method, but a stage in the post-melting solidification is datable. At some point during solidification after general internal melting a particular daughter element ceased to be able to move away from the actual site of the radioactive decay of the parent element—in other words the system ceased to open for this daughter element. This is the point in the parent body history measured by the solidification age method.

The decay of the following unstable nuclides is utilised:

$$^{87}Rb \rightarrow\ ^{87}Sr \quad . \quad . \quad . \qquad \text{half-life } 47 \times 10^{10} \text{ years}$$
$$^{187}Re \rightarrow\ ^{187}Os \quad . \quad . \quad . \qquad \text{half-life } 43 \times 10^{9} \text{ years}$$
$$\left.\begin{array}{l} ^{235}U \rightarrow\ ^{207}Pb \\ ^{238}U \rightarrow\ ^{206}Pb \end{array}\right\} \text{combined in } \left\{\begin{array}{l} \text{half-life } 0{\cdot}71 \times 10^{9} \text{ years} \\ \text{half-life } 4{\cdot}51 \times 10^{9} \text{ years.} \end{array}\right.$$

with $^{207}Pb/^{206}Pb$

Rubidium–Strontium Method

This method is the most successful for determination of solidification ages of meteorites, besides being widely applied to terrestrial rocks. Originally based on $^{87}Rb/^{87}Sr$, it was later refined with the incorporation of a second ratio, $^{87}Sr/^{86}Sr$. The two ratios are plotted together for a number of specimens of different composition, ideally producing a straight-line plot which indicates the radiometric age by virtue of its angle of slope.

As applied to meteorites, the fact that unfractionated meteoritic material may exist must be considered in applying this method: such material may well occur among the meteorites measured. There is no certainty that all meteorites represent material fractionated during the melting of the parent body. The method involves the use of a number of meteorite samples in conjunction to give a sloping straight-line plot or 'isochron', and any unfractionated material would give a spurious point on the isochron. We can, however, be certain that achondrites are significantly geochemically fractionated (possibly by a selective volatilisation), and some chondrites to a lesser extent, but most chondrites yield results that leave some doubt whether they are fractionated.

225

Despite these doubts, the method yields results around 4.37×10^9 years, a value in accord with those obtained from other methods.

Rhenium–Osmium Method

Very similar to the rubidium–strontium method, this is, however, less satisfactory. The daughter element has limited mobility, which leaves doubt concerning the basic assumption on which the method is based. The parent and daughter element are not susceptible to geochemical fractionation, a requirement of the method. Results on iron meteorites give an isochron based on $^{187}Re/^{186}Os$ and $^{187}Os/^{186}Os$ ratios indicating a solidification age of $4.0 \pm 0.8 \times 10^9$ years. The results show a wider variation, individually, than in the case of the previous method, and must be viewed with reserve.

Lead/Lead Method

The advantage of this method is the fact that the decay constants are determined with a high degree of accuracy. Also, the measurement of an isotopic ratio for two isotopes of the same element is, in practice, simpler. But the presence of non-radiogenic lead in all meteorites to some extent counters these inbuilt advantages. The method requires the use of two meteorites in conjunction, assuming the same lead and uranium isotopic composition for both, and the same solidification age. For achondrites and chondrites, including carbonaceous chondrites, an isochron based on $^{207}Pb/^{204}Pb$ and $^{206}Pb/^{204}Pb$ gives a solidification age of $4.6 \pm 0.1 \times 10^9$ years. There are certain difficulties in applying this method to iron meteorites, but these also tend to give values just above the 4.6×10^9 isochron. A more complicated fractionation history is suggested for the calcium-rich achondrite Nuevo Laredo, Mexico, for, whereas irons yield ages suggestive of only simple metal/silicate phase fractionation, the results for this meteorite suggest that it first lost its nickel and other siderophile elements, then had its silicate phase enriched in calcium, aluminium, uranium, and thorium with depletion in magnesium, sodium, and the heavier alkalis, finally losing the chalcophile elements thallium, lead, and bismuth.

There is evidence that the period of meteorite synthesis was short —that is the time span from overall melting of the parent body to completion of fractionation. There is also evidence stemming from this method that chondrites were appreciably fractionated as regards lead, at least, and this suggests that the uncertainties inherent in the Rb/Sr method may be of no account.

These three methods suggest that the solidification age of the parent body (the termination of melting) took place about 4.6×10^9 years ago.

Gas Retention Ages

Noble gases result from the decay of certain long-lived radionuclides:

$$^{40}K \rightarrow {}^{40}Ar$$
$$^{238}U \rightarrow {}^{206}Pb + {}^4He$$
$$^{235}U \rightarrow {}^{207}Pb + 7{}^4He$$
$$^{232}Th \rightarrow {}^{208}Pb + 6{}^4He.$$

These noble gases only remain in situ at low temperatures, being able to move through mineral lattices at raised temperatures. They can thus be used to date the last heating of meteoritic material (other than ablation heating which affects only the surface area). The age obtained should represent either the date of termination of cooling of the parent body, or of any subsequent heating of the meteoroid after separation from it, while orbiting in Space prior to arrival on Earth.

Potassium–Argon Method

Problems arise on account of the low potassium content of meteorites and also the difficulty of extraction and measurement of argon. There is also uncertainty concerning the radiogenic ^{40}Ar and cosmogenic ^{40}K content of meteorites.

Uranium–Helium Method

This method cannot be used in the case of irons, because most of their helium content is cosmogenic, not radiogenic. Chondrites and especially calcium-rich achondrites do, however, form suitable subjects for such measurements.

Both methods give a wider range of results than the previous methods. Ideally, these results could be interpreted as indicating formation of the mineral phases comprising the meteorite after a high-temperature episode $4.6 \pm 0.2 \times 10^9$ years ago. However, when two components of a polymictic breccia are dated separately (Kunashak, CHy, and Pervomaiskii, CHy, both from the USSR, have been studied in this way) the ages obtained show discordance, in the case of both methods. Thus, a more complex model must be sought:

227

among those suggested are:

(a) Continuous diffusion loss at not very high temperature during cooling of the parent body.

(b) Some diffusion loss at not very high temperature during storage in parent bodies after the main cooling.

(c) Solar heating of meteoroids after break-up of the parent body.

Temperatures of as low as 300 °K are critical for gas retention. Taking the first model above, the results would indicate, because of the small scale of the disparities between gas retention and solidification ages, that a small parent body a few hundred kilometres in diameter must have cooled rapidly to below 300 °K. Or else, the chondrites which show such low discrepancies must have come from the rapidly cooled surface zone of larger bodies that cooled more slowly in their interior. Another possibility is that break-up of the parent bodies was multistage, and what the gas-retention ages indicate is a certain stage in this multistage break-up, break-up to metre-size fragments following the final loss of noble gases, and cosmic-ray exposure then commencing.

The second model invoking late parent body storage could well account for all the anomalies. It is compatible with the calculated cooling period for an asteroid of 700 km radius, about 700 million years. A certain amount of solar heat or heat derived from short-lived radioactivity might have maintained a sufficient temperature for some further noble gas diffusion and loss to occur.

The third model refers the further loss of noble gases to the cosmic-ray exposure period—the meteoroid stage. The high degree of orbital eccentricity required for meteoroids to arrive on Earth at all would result in solar heating at perihelion. There has been some support for loss at this late stage of noble gases, in the form of predictions of the amount the gas retention age will fall below the solidification age, based on cosmic-ray exposure ages. Such predictions have been very close to the subsequent experimental results. The value for the Pribram chondrite, of known orbital characteristics, has been predicted from consideration of its orbit and cosmic-ray exposure age, and again there is a close agreement with the experimental results.

In deciding whether the main cooling in the parent body, late storage in the parent body, or loss in the meteoroid stage are the factors responsible for the gas retention/solidification age discrepancy and the polymict breccia discrepancy, many other factors need to be considered. For instance, the evidence from $^{128}I \rightarrow {}^{129}Xe$ decay (considered below) suggests that the cooling period of the parent body was long. Further, even taking parent bodies of large asteroid size, the internal heating would be likely to have been considerable. The possibility of a systematic difference in gas retention ages, between recrystallised chondrites (low) and normal chondrites (higher), has also been suggested. Also, significant is the fact that the shock-modified shergottites, numbering only two, yield identical potassium–argon and uranium–helium ages. This strange equivalence suggests that meteorite collision rather than parent body processes may be the factor explaining the gas-retention age anomalies. Contrasting gas-diffusion rates and half-lives of radionuclides would otherwise be expected to preclude such an equivalence of ages. It is possible that bodies up to 100 km radius separated from larger parent bodies and were fragmented again by secondary collision, resulting in a sudden internal temperature drop: the meteoroids of 'metre' size, familiar to us, being then produced. However, there is conflicting evidence of discrepant potassium–argon and uranium–helium ages suggesting that reheating actually occurred during the cosmic-ray exposure (meteoroid) stage.

The exact nature of the Kunashak and Pervomaiskii polymict breccias needs clarifying. If, as seems likely, these are light-dark breccias, it must be noted that modern interpretations of such breccias are not as polymict breccias in the true sense, but as associations of shocked and unshocked material of similar origin. If this is the case the anomalies in a gas-retention age may well be related to shock-induced loss of noble gases. Then, the question must be asked, how much of the visible shock effects in meteorites are due to meteoroid collisions during the cosmic-ray exposure stage? The eucrite enclaves within the Mt Padbury mesosiderite show what appears to be the incipient development of the shock vitrification (maskelynitisation of feldspars) that produced the shergottites. Are not the light-dark breccias and shergottites products of late, parent body shock effects? Was not the late noble gas loss partly so produced and partly produced in the meteoroid stage?

Extinct Radioactivity

In theory, radionuclides with half-lives of 10^8 years or less must once have been present in the solar system, providing nucleosynthesis took place at no long interval after its formation. A method of age dating based on this premise is extensively used. It involves several assumptions concerning the isotopic homogeneity of the solar nebula, and the acceptance of one of three possible models for nucleosynthesis:

Single sudden event.
Continuous nucleosynthesis throughout the galaxy.
Both galactic and local (solar nebula) nucleosynthesis.

The values determined by this method are not true ages. Measurements are forwards in time from an elusive starting point, not backwards in time to a specific event. The ages obtained are called decay intervals. The unsupported decay of say ^{129}I is measured without chemical fractionation and without retention of the decay product. The methods involve the decay of:

$$^{129}I \rightarrow {}^{129}Xe \quad . \quad . \quad . \quad \text{half-life 16·4 million years}$$
$$^{244}Pu \rightarrow {}^{238}U \quad . \quad . \quad . \quad \text{half-life 76 million years.}$$

The onset of xenon retention in the meteorite-to-be apparently occurred at quite widely different times in various meteorites. The fact that the retention of xenon apparently requires a cooling to below 300 °K means that the variable results can be interpreted in two ways:

(a) The formation of the meteorite-to-be by nucleosynthesis occurred at different times in the case of different meteorites.

(b) The meteorites come from different parts of the parent body, and those from the outer zones cooled first and therefore will give high decay intervals, while those from further in towards the centre will have cooled later and should give low decay intervals.

Another factor could be the size of the parent body. If there were several parent bodies of asteroidal size, not only will disparity in decay intervals be produced by virtue of different depths within the parent body, but also there will be differences due to the different sizes of the parent bodies—for example a larger parent body will be expected to produce core material cooled more slowly, and, therefore, even if formation of the meteorites-to-be by nucleosynthesis actually occurred simultaneously, meteorite material from its extreme core will show a lower decay interval than that from a smaller parent body.

230

Other complications could be introduced by virtue of complications during accretion prior to the cooling to the point at which ^{129}Xe could be produced, and it is clear that all the scientist can do is gradually progress, as more and more results introduce further complications and exclude certain models, towards a really satisfactory model of the history of the early solar system, from nucleosynthesis through accretion to cooling, initiation of decay, and final extinction of the parent radionuclide.

In spite of the uncertainties of the method, ^{129}Xe measurements have given us some valuable clues. The decay intervals are generally consistent with a radius of about 200 km for the parent body or bodies (that is asteroidal rather than planetary size) and two independent arguments have also suggested this. The decay intervals from the stony meteorites do seem to be compatible with enstatite chondrites being situated further out than common chondrites, possibly with very compacted chondrite material coming from deeper than other common chondrites, and irons coming from the core of the parent body. The results from the Murray, Kentucky, carbonaceous chondrite suggest a long cooling time, yet the presence of primordial noble gases suggests the converse. This anomaly has been explained as being due to radiogenic noble gases being situated in sites of less retentivity than the primordial noble gases. This is possible, as the radiogenic product may well have remained in the mineral phases that carry the iodine parent (though this is not certain, for it may have diffused to new sites). In other words, the radiogenic noble gas requires only slight heating to be moved out of the system, whereas the primordial gases would require a much stronger heating.

It is not certain that the ^{129}I/^{129}Xe decay took place in the solid meteorite-to-be (ie, in the parent body) and not in the solar nebula before the formation of any solar bodies, though, if this was the case, the difference in the isotopic mass spectra of meteoritic and terrestrial xenon must be explained by some form of early chemical fractionation in the nebula, which must have been unevenly mixed. Much discussion of the two alternatives has been published, and experiments devised to test which is correct. Though the evidence is not conclusive, it strongly favours decay in the solid, already accreted parent body rather than the nebula. Also, from experiments that involve measurement of the point during heating up of the meteorite mass at which the various noble gas isotopes are released has come evidence that the ^{129}Xe does reside in the mineral phase that contains the parent

231

[129]I. In the case of the enstatite chondrites, this is believed to be daubreelite, or another sulphide, for iodine is geochemically chalcophile.

Plutonium-244

Excesses of heavy xenon isotopes in terrestrial xenon have been attributed to possible fusion of a heavy radionuclide such as ^{238}U or ^{242}Cm. Lately ^{244}Pu has been suggested. Considerable argument has developed concerning the presence of these xenon isotopes in minute quantities in meteorite material.

The results are supposed to be compatible with nucleosynthesis occurring in a well-stirred galaxy, so that the initial ratio of any two nuclides was determined in the solar system only at the stage of isolation of the material from this well-stirred mix. Reynolds, however, finds this conclusion to be insecurely founded, and suggests that much more data are needed before such a conclusion can be drawn.

Other Extinct Radioactivities

Other extinct radioactivities such as ^{205}Pb/^{205}Tl and ^{107}Pd/^{107}Ag have been investigated. Both measure the cessation of melting, that is the solidification, of the parent body after accretion, not cooling to a relatively low temperature subsequent to the solidification as is the case with the methods based on measurements of xenon isotopes. This introduces difficulties in considering decay intervals obtained from these methods and the previous methods alongside one another. Out of these studies certain evidence has, however, come—there is some reason to accept very rapid accretion, melting, and solidification of the parent bodies, followed by a rather longer cooling period. There is some evidence favouring local nuclear processes just before or during the formation of the solar system, quite apart from the main galactic nucleosynthesis—charged particle acceleration during the dissipation of the Sun's magnetic field has been suggested as the agency.

Isotopic Spectral Discrepancies Between Terrestrial and Meteoritic Matter

The differences between isotopic spectra of meteoritic and terrestrial matter are attributed by Goles and Anders to charged-particle-induced spallation within the solar nebula, meteoritic material being supposedly derived from lower density matter in the asteroid belt, and thus more affected by this process than terrestrial matter. This explanation is preferred to others advanced by Cameron and by Krummenacher *et al.*, because the non-volatile element barium shows similar isotopic anomalies to xenon, which is a gas. The discrepancies seem to be due to a process which affected meteoritic material rather than terrestrial material, and not the converse. A process involving intense proton and thermal neutron irradiation of meteor-sized planetesimals has been suggested by Fowler *et al.* as the actual process involved.

The introduction of new techniques of measurement is gradually allowing revision of isotopic spectra for many non-volatile elements formerly considered to show no discrepancy between meteoritic and terrestrial matter. With more evidence of this sort it may become possible to draw up a much more exact model of the early history of the solar system: at present the extinct radioactivity methods provide fascinating glimpses into this far-off time, but the assumptions involved, the inaccuracies involved in measurement, the possibility of interference by other radioactivity processes, and the number of models still tenable, all of which have something in their favour and much against them, are such that this remains a very open field of research.

Primordial Noble Gases

Unusual quantities of noble gases, first discovered in the Pesyanoe enstatite achondrite, were later located in carbonaceous chondrites, common chondrites, calcium-rich achondrites, and ureilites. These gases have an isotopic ratio pattern that is quite unusual, and are not due to atmospheric contamination, nor do they appear to be radiogenic. It has been suggested that they represent a retention of an abnormal amount of primordial noble gases, from the nebula. Possible retention in silicate minerals possessed of gas-trapping structures does not appear to explain this phenomenon. Sealing within the asteroidal parent body as an internal atmosphere by the effect of a permafrost surface has been suggested: this could fit in with the 200 km radius dimensions favoured by certain evidence. It has also been suggested that an internal sintered zone could assist in the retention of these gases—such a sintered zone is favoured in a model for the parent body suggested by Goles and Anders. Uneven distribution of these gases in the Pantar meteorite, a chondrite showing a breccia texture with the familiar dark and light areas of which the dark areas appear to have suffered shock, has led Konig *et al.* to consider the possibility that migration and trapping of this gas fraction in the dark areas, preferentially, is a shock effect. Evidence from the Abee enstatite chondrite, obtained by Jeffery and Reynolds, suggests, however, that in that case, the primordial gases are trapped in a particular crystal lattice. The amount of these gases retained by these meteorites is greater than that retained by the Earth, and individual noble gases show varying degrees of retention. What loss occurred of such a relatively heavy gas as neon must have been from a much lower gravitational field than that of the Earth at the present time—neon loss is now impossible—and Suess has suggested that, in the case of the Earth, an escape occurred in the planetesimal stage before accretion: or, alternatively, during an era when the Earth's period of rotation was so short that centrifugal force largely offset gravitational force in the upper atmosphere. The meteoritic ratios are not, however, compatible with such explanations. What is needed, to explain these ratios, is a process that could give the values for neon without producing large elemental fractionations between neon and helium. A diffusion mechanism has been found able to account for the observed values for meteorites, and also for the Earth.

Alternative propositions that these noble gases are secondary accretions have come from Urey and Cameron. The former suggested a source in cometary material while the latter suggested that much of the Earth's xenon comes from the Sun. Anders concludes that the diffusion mechanism could still apply to noble gases from such sources. He notes that it is difficult to achieve any greater certainty concerning the detailed history of the primordial gases in the meteorites, but their retention in achondrites does suggest an early stage trapping by solid particles prior to any elemental fractionation, accretion of these particles under low temperature conditions (otherwise, the noble gases would have been driven out), and late movement into the achondrite material by diffusion (after the high temperature process of cooling and crystallisation from a melt that formed the achondrite material). He concludes that the noble gas history of chondrites is less well understood, that it is not even certain whether the fractionations were in parent body or the nebula in this case. He notes that meteoritic histories are likely to have been such as to allow several opportunities for diffusion of gases under favourable temperature gradients, and more than one process may have been involved. He regards the lack of neon/helium fractionation as all the more extraordinary, in view of these 'diverse opportunities'.

234

19: The Origin of Meteorites

We are, at least, certain that meteorites come to us from out of the Sky—indeed, they are defined as material that falls to Earth from outer Space. Yet, secure as we are in even this knowledge, we must exert care before we apply the term meteoritic to all material that possesses 'meteoritic character'. For we cannot discount entirely the possibility that material not readily distinguishable from meteoritic material may have been brought up, from time to time, from the depths of the interior of our own planet. Cryptovolcanic and diatreme activity seem to offer some possibility of effecting just this. Kimberlite pipes, besides bringing up diamonds, are diatremes that bring up solid chunks of igneous rock believed to be derived from the Earth's main layer or 'mantle', and other types of diatreme are believed to have done the same. We also have the uncomfortable and inescapable fact that lumps of nickel-iron, of low nickel content and so displaying no organised etch pattern, are contained, alongside disseminated specks of the same material, in basalts of volcanic origin on Disko Island, Greenland—the Disko Irons first described by Nordenskjold and Bøggyld. Though supposed terrestrial analogues of chondritic material remain unconvincing, and there is no certain terrestrial occurrence of such material, there remains the possibility of meteoritic iron being simulated by material of terrestrial origin.

It is worthy of note that geochemists and geophysicists have long relied on analogies between the Earth's 'mantle' and chondrites, and the Earth's 'core' and meteoritic iron, and continue to refer to the 'chondritic' Earth model: yet nothing like chondritic material has ever been brought up by deep seated diatreme eruptivity, even though the nodules of supposed upper mantle origin are mostly peridotitic, and, so compositionally, if not texturally, very similar to chondritic material.

Another interesting analogy is that drawn between the calcium-

rich achondrite meteoritic material and terrestrial dolerites or gabbros of igneous intrusive origin. In the case of the eucrites the
resemblance is uncanny, but we have seen very similar material,
though compositionally different because of its high titanium content, in lunar samples brought back by Apollo 11, 12, and 14, and it
is clear that basaltic magmas crystallised in a very similar manner
within the very small asteroidal parent body to the meteorites, within
the rather larger Moon, and within the planet-sized Earth—the
physical effects of small or large global dimensions of the particular
solar system member do not seem to have been so productive of
textural variations as might be supposed.

Geochemists and geophysicists nowadays are regarding meteoritic
material as products of solidification in *low*, not *high*, pressure environments: the presence in meteorites of the silica polymorph
tridymite which is not stable under high pressures seems to rule out a
high-pressure environment, and the low-pressure environment is
much more compatible with recent indications of the probable size of
the meteorite parent bodies. With this abandonment of the concept
of a high-pressure environment, comes abandonment of the idea that
very high pressure was an essential requirement for the formation of
the coarse Widmanstätten alloy patterns of octahedrites. Several
parent bodies are also now widely favoured, the largest no greater
than Ceres (770 km diameter, the largest of the asteroids), and some
possibly much smaller. The chondritic Earth model has become somewhat superseded, for in contrast to these small parent bodies, our
own Earth must have a high-pressure mantle zone and the deeper
mantle material must, apparently, consist of dense mineral polymorphs quite unfamiliar to us, possessing much more compressed
crystal lattice structure than the olivines and pyroxenes familiar to us
in chondrites and superficial terrestrial igneous rocks. And, if the
chondritic Earth model seems to be largely superseded (except in the
sense of chondrities and their chemical relation to the average composition of the solar nebula), so, perhaps, should the nickel-iron
model for the Earth's core, for this too was based on the analogy
with the meteoritic iron! This analogy was, of course, based on the
premise that Widmanstätten alloy patterns were of extreme high-
pressure derivation, and thus an analogy could be reasonably drawn
with the Earth's core, which must be under extremely high hydrostatic pressure. However, strange as it may seem, whereas the
chondritic Earth model remains only correct in that the Earth's upper

mantle and chondrites are both essentially of peridotitic composition, the core analogy may yet prove to be true. It fits in very well with what we know of the geophysics and geochemistry of the Earth, and the existence of the Earth's complex and variable magnetic field, and, to date, though other models have been proposed, no really satisfactory alternative model has appeared. But, if we continue to accept the concept of a nickel-iron core, we must not retain the direct analogy with meteoritic alloy patterns or compositions. Ringwood would like to add a seasoning of silicon to the terrestrial core alloy, and silicon does partake in the metal alloy of some meteorites (for example, Horse Creek and Mt Egerton). This addition would, he believes, satisfy the geophysical requirements better but, in truth, we can still no more than hazard a guess at the exact composition of the Earth's core—it may have unguessed at minor constituents in the metallic alloy. And, as Anders (1971) notes, the meteorites reflect in their composition a progressive development of metal/silicate fractionation; and there is evidence in the contrast between planetary densities, and in the contrast between the chemistry of terrestrial, lunar and meteoritic material, that the various planets, satellites, and asteroids are products of quite different individual fractionations.

The acceptance of the low-pressure environment for meteoritic material has perhaps made simulation by material brought up in diatremes less likely—but we cannot yet entirely discount terrestrial nickel-iron alloys of a character liable to confusion with meteoritic iron. It is probable that studies of germanium and gallium contents, or of cosmic-ray induced isotopes, could be extended to some of the 'meteoritic' iron occurrences associated with craters such as Cañon Diablo and Wolf Creek (Chapter 20) concerning which some ambiguous evidence has been cited in arguments questioning their meteoritic impact origin. Though the evidence for meteoritic origin is very strong, an iota of doubt still lingers in the mind—the first crater displays some very odd relationships to the host terrain (noted by Hager): and is situated not far distant from cryptovolcanic structures and kimberlite pipes, leucitic diatremes, etc. The second is situated at the south-east end of a linear group of leucitic (lamproitic) diatremes including the cryptovolcanic structure of Mt Abott, now known to contain kimberlite bodies, and also has an anomalous structure. It is odd that two of the world's largest Cainozoic-Quaternary meteorite impacts should have occurred very close to kimberlite/leucitite igneous eruptive provinces, for, except for the Siberian and South

African Provinces, such provinces are geographically of extreme rarity and occupy very small areas of the globe.

Hypotheses of Meteorite Provenance

Hypotheses of meteorite provenance have, indeed, included some that invoked terrestrial eruptive ejection and return to Earth! Among the more significant hypotheses advanced during the last century, may be cited:

1 Pre-Chladni (continued to be accepted widely in France up to 1803, and restated by Mukhin in 1819). Atmospheric origin: condensed during thunderstorms from dust, vapour, etc.

2 Chladni (1790). Solid debris from interplanetary space, derived from external impact on, or internal explosion of, larger bodies, resulting in fragmentation.

3 Laplace (1827). Volcanic eruptions on the Moon.

4 Ball (1913). Ejection from terrestrial volcanoes into Space, during episodes of more violent eruptivity than is known today, in ancient times.

5 Lodochnikov (1939). Revived the same theory.

6 Olbers (1840) Fragments of disintegrated planets (sic) similar to asteroids, which he discovered.

7 Schiaparelli (1910) Fragments of disintegrated comets.

8 Kulik (1942). Advocated a similar origin.

9 Vernadskii (1945). Advocated an origin as products of interstellar Space, entering the solar system on hyperbolic orbits. Vernadskii, in advocating a galactic provenance, dismissed planetary (=asteroidal) provenance as 'an assumption based on seventeenth-century ideas, alien to celestial mechanics and universal contemporary views'. This resounding establishmentarian rejection would find little support today, but it draws attention to the fact that the now favoured model, derived from our modern technology and innumerable lines of evidence, does closely match that of the seventeenth-century (eighteenth-century in the English sense) scientist, Chladni, derived from very lucid interpretation of very limited evidence! It does not in fact contradict the conclusions of Diogenes of Apollonia!

10 Astapovich (1939). Derivation in part from without and in part from within the solar system was favoured by this Russian scientist. He believed that catastrophic ejection of magma and

solid fragments takes place outside the solar system, possibly during immense cataclysms of bodies of the type known as 'dark stellar companions'.

11 LaPaz. Almost alone amongst living scientists has maintained that meteorites come from beyond the limits of the solar system on hyperbolic paths: echoing the beliefs of Vernadskii.

12 Urey. Believed that some meteorites come from the asteroidal belt, whereas others are spalled off the Moon by the agency of impacts. A variant of this idea invokes mixed asteroidal and lunar origin for mesosiderites.

Among the most recent attempts to formulate a model for the origin of the meteorites is that of Mason (1967), and this model appears as realistic as any such model, though there are objections to it, as there are to any model based on our present state of knowledge. For one of the contradictions of meteoritics is that, whereas in almost no scientific subdiscipline has detailed knowledge accrued so rapidly in the last twenty years, yet there are profound gaps remaining in the availability of evidence that prevents any rigid exclusion of all but one model.

The Story of Meteorites

The story of meteorites is believed to start with the formation of the solar nebula by nucleosynthesis, probably during the course of a supernova event—an outburst in which a star seems to blow most of its material cataclysmically outwards into Space, never returning to its former state. Through a stage when the matter had the form of interstellar dust and gas, a further stage was reached in which this material formed the primitive material of the solar nebula. Accretion produced the separate planetary bodies and the asteroids. Mason rejects the theory, long advocated by earlier scientists, of disruption of a single planet intervening between Mars and Jupiter to form the present asteroidal belt: he believes that its very simplicity has become quite irreconcilable with the increasingly complex history of the meteorites and their parent bodies revealed by fast accruing new data. The geologist, Daly, actually deduced a radius (3,000 km), a volume ($0 \cdot 1 \times$ that of the Earth), a mass ($0 \cdot 07 \times$ that of the Earth), a mean density ($3 \cdot 77$), and a diameter for the nickel-iron core (1,000 km), for this hypothetical, lost planet, which has been referred to as Phaeton (after a legendary character who rode his fiery chariot across the skies). Despite the poetry of this name, Phaeton, like the planet Vulcan (Mercury's supposed sunward brother), must probably now be relegated to the limbo of planets that never were.

Modern evidence favours several meteorite parent bodies, but attempts to fix the exact number come headlong up against conflicts of evidence. The main difficulty is that there are a number of discontinuities in the progressive range of meteoritic properties, but one can never be certain whether a particular discontinuity applies a single parent body or represents a contrast between different and dissimilar parent bodies. There are, in fact, many such discontinuities evident in ranges of igneous rocks crystallised from magmas, and these must, of course, be regarded as inherent characteristics of certain physico-chemical processes, not due to provenance from separate worlds! Study of one particular geochemical or geophysical property—for example gallium and germanium contents in meteoritic irons—may give you a nice, neat model of parent bodies of different sizes, but such a model may not stand up to the rigorous test afforded by the evidence derived from meteoritic polymict breccia associations. At the present stage of knowledge, there seems not to be a high degree of agreement between models derived from different lines of research, and it does not seem that even the single parent body theory can as yet be completely discounted, even if it is out of favour. For the possibility that metal areas in the parent body or bodies had raisin-bread distribution rather than forming a single central core seems to leave the door just ajar for a revival of this theory.

Mason regards carbonaceous chondrites as samples of the non-volatile matter of the solar system. He regards the group of chondrites as a whole as reflecting fractionation processes within the solar nebula prior to parent body accretion. Overprinting these initial variations are effects supposed to reflect contrasting accretion environments, contrasting physico-chemical conditions within the parent body prior to disruption, and contrasting parent body dimensions.

He considers that

1 birth of the solar nebula
2 formation of chondrules
3 accumulation of the parent body
4 differentiation of the non-chondritic meteorites within the parent body

all occurred within a relatively short space of cosmic time, an interval many orders shorter than the 3,500 million years covered by the surface geological record of our Earth. The various radioactivity clocks mentioned in Chapter 18 suggest that this short time interval occurred about 4,700 million years before the present. Mason suggests that the parent bodies ranged from that of the largest asteroid, Ceres (770 km diameter), down to 'football size'—presumably meaning that some parent bodies had diameters of but a few metres. There is very good evidence for large asteroid-sized parent bodies in cooling rates determined for iron meteorite alloys, and taken to be proportional to the size. After accretion the parent bodies heated up and melted internally by the agency of radioactivity and other causes such as gravitational and internal tidal effects, and the cooling rate of the larger, partly internally melted parent bodies will have been proportionally slower. The pallasites appear to be fragments of a large parent body, about the size of Ceres (assuming they do come from a concentric-shelled body).

We have a maximum possible size for the parent body or bodies, and a larger planetary parent seems ruled out, unless the supposed contrasting dimensions for individual parent bodies indicated by cooling rate studies actually apply to cooling zones within a single 'raisin-bread' structured parent body of planetary dimensions, not a concentric zoned parent body. Evidence from isotopic studies given in Chapter 18 also favours large asteroid-sized parent bodies.

Mason seems to be on less firm ground when postulating numerous small parent bodies ('football size') for the chondrites, in contrast to the few large asteroid-sized parent bodies that he accepts for the irons and achondrites. The evidence of the Bencubbin polymict breccia given in Chapter 14 seems, in particular, quite incompatible with this model, which seems to have been derived in order to explain anomalous exposure 'ages' obtained for chondrites (Chapter 18). Another resolution must, it seems, be sought for this anomaly.

The break-up of the parent bodies, whatever their number and dimensions, and the projection of the fragments so derived, the *meteoroids*, into eccentric orbits, including some that pass through the orbits of Mars and the Earth to perihelia between Earth and Sun, remains unexplained. Only a minority of the vast population of meteoroids would have been thrown into such special orbits that Earth impact was at all possible. Thus, we are only sampling a very small fraction of the meteoroid population in the solar system, and any geophysical or geochemical conclusions drawn from study of this special sample must be accepted as based on incomplete evidence.

The author has suggested that the break-up of the meteorite parent body may have been the result of degassing—the small planet and satellite bodies, the Moon and Mars, are intensely cratered, and if we diverge from the popular belief of American scientists and dare to attribute the giant craters and maria to endogenous causes, consequent on a grand-scale eruptivity accompanying early degassing, then we can envisage these smaller globes as having been nearly broken apart during the formation of the largest maria. The even smaller asteroidal globes may have actually suffered disintegration, being fragmented by pressures from within, during an early, even more abrupt degassing. Just because there is little water in meteoritic material and very little, also, in known lunar material, we do not have to rule out gases in considering eruptivity affecting the smaller globes of the solar system. Even allowing this possible answer to the age-long problem of the cause of disintegration of the meteorite parent body, the manner of projection into such diverse orbits remains quite unexplained, and projection into a Mercury-passing orbit such as that of Icarus may seem difficult to attribute to purely internal causes. This idea is indeed entirely empirical and may well prove to be untenable in terms of rigid physical analysis, but such ideas are badly needed to fill this profound and fundamental gap in meteoritic theory. This idea incidentally seems to be quite incompatible with Mason's concept of 'football-sized' parent bodies, for it relates to large asteroid-sized parent bodies.

Whatever its origin, and whatever the size of the parent body or bodies, and their number, such a break-up seems to have occurred. It is fundamental to all now widely accepted theories of meteorite genesis, other than those of favouring a provenance outside the solar system. It caused the meteoroids, now identities in their own right within the complex array of members of the solar system, to orbit for immense periods in Space. Whether during this period of meteoroid orbiting they were subject to repeated collisions, each collision tending to further reduction in size of the masses as some authorities suppose, or whether, as seems indicated by some evidence, such collisions were relatively rare, remains an open question. Cosmic rays are supposed to have attacked the surfaces of meteoroids, and such collisions to have laid bare new surfaces for onset of attack by cosmic rays. Mason suggests that the cosmic-ray exposure age anomalies can be

242

reconciled with two asteroids colliding 520 million years ago, one being of olivine-hypersthene chondrite composition and the other of octahedrite composition, this collision being followed by later collisions between small, fragile, easily disrupted chondrite masses; but other evidence seems to conflict with such a resolution of the problem, and one is justified in wondering if there is not some quite unsuspected reason for the cosmic-ray exposure age anomaly, quite apart from modification by repeated meteoroid collisions. Solar heating at perihelion also may have affected the meteoroids. The attribution by Mason of chondrule formation to a pre-accretion stage leans heavily on the suggestions of Wood. The author finds it very difficult to accept that chondrules were formed prior to the accretion of the parent body, believing them rather to be due to a peculiar form of spheroidal growth in a quasi-magmatic stage, within the parent body, under low 'hydrostatic' pressure and in the presence of a volatile atmosphere. Both Anders and Ringwood have also, on occasion, tended to favour production of chondrules by internal, quasi-volcanic processes. Mueller refers the production of chondrules to the actual accretion stage, but does not seem to afford the chondrules pre-accretion existence as discrete, spherical identities. Anders referred the chondrules to the main shell of his hypothetical parent body, but Ringwood does not infer such a regular zonal arrangement. There can be little doubt that chondrules, with their magmatic textures and part glassy or holocrystalline character, and in some cases indications of supercooling, now reflect rapid crystallisation. The real question is, 'Was this superimposed on a spherical body that had its origins in the pre-accretion stage, as a planetesimal within the nebula, or was this the actual process of formation of the spherical body, which had no prior existence?' The author believes that the latter alternative is the correct one.

The status of the carbonaceous chondrites may not be quite that suggested by Mason. It may be that they represent volatile-retention or volatile-concentration areas patchily distributed in the parent body, as has been suggested in Chapter 14. Mason believes they represent the material situated at sites of accumulation at low temperatures located at the periphery of the solar nebula, where gas was more abundant than solid material and accretion of large parent bodies, with consequent interior heating up not possible. However, their rarity and insignificant size and the nature of their known occurrence as enclaves in polymict breccias, *within thermally recrystallised material*, suggests that this is to give a far too broad cosmic dimension to what are, in truth, no more than a handful of small lumps and a few minute enclaves in breccias, representing a very minute fraction of the very extensive spectrum of meteoritic material and occurrences. The truth may well be that they are derived from unmodified residual patches in a generally modified meteorite mass, patches that contain some products of crystallisation at quite high temperatures, in the form of silicates and metal, and some later condensed material that has never suffered heating after condensation, but has tended to vein and invade the high-temperature material during minor relocations within the parent body, and so form the mechanical mixture that we see today.

Summary of Conclusions

The author would modify Mason's model as follows:

1 Chondrules are likely to be of parent body accretion stage or post-accretion eruptivity origin.

2 Carbonaceous chondrites are, perhaps, something of a red herring, representing only minor areas of anomalous volatile retention and unexplained resistance to thermal modification, despite their extreme geochemical significance.

3 Parent bodies may have been multiple: if concentrically zoned around a core, they are unlikely to have exceeded large asteroid size. Numerous 'football-sized' parent bodies seem unlikely, and chondrites probably come from the same parent bodies as the non-chondritic meteorites, though, if so, the various zones in the parent body had abrupt discontinuities. Alternatively, a somewhat larger, single parent body possessed of many raisin-bread metallic foci can be entertained as a remote possibility.

4 There was probably a sort of shell configuration, but it was almost certainly crude, and all zones were heterogeneous, tending to patchy and breccia character. Individual patches may have behaved as closed systems, there being little reaction or transfer of material across the boundaries.

5 (a) The chondrites of all types probably represent an outer zone: the stony-irons, achondrites, and some iron material probably represent a deeper zone, with the irons forming the core zone.

5 (b) Such zonation could, again, well have been around metal-cored foci in a raisin-bread structured parent body.

244

6 Abrupt degassing may have been the agency that caused the break-up of the parent body.

7 Collisions in the meteoroid stage have probably been over-emphasised: they may have been relatively rare. Solar heating at perihelion is a likely complication of the meteoroid stage.

The author thus favours a modified asteroidal theory. It is difficult to support the theory of LaPaz, deriving meteorites from interstellar Space, beyond the limits of the solar system. The evidence seems to be overwhelmingly against hyperbolic paths, though some sort of a case can be made for modification of all originally hyperbolic trajectories by solar system perturbations. And, if we reject provenance in interstellar Space, where else but the asteroids can one see a source for meteoritic matter? The Earth/Moon system seems ruled out: the Earth has a high-pressure internal environment: lunar material is quite unlike meteoritic in elemental composition. The planets and their moons and the comets remain as the only other possible sources. There is little reason to favour the planets or their moons as a source, and the comets, though more likely sources, seem to be masses of quite dissimilar character, not essentially lithic or metallic, whereas asteroids are known to be formed of solid, commonly ir-regular lumps of rock or metal (or both mixed), and would seem to be the obvious source, even if the evidence were not strongly in favour of them. Yet, some authorities have lately turned towards the pos-sibility of a source in periodic comets, for at least some carbonaceous meteorites, and we cannot yet claim to have any certainty, at this state of knowledge, of the origins and source of meteorites. We can but say that asteroidal provenance seems overwhelmingly favoured by the evidence.

20: Possible Meteoritic Origin of Large Cryptic Craters and Crypto-explosion Structures

In Chapter 4 small fragmentation craters were described, structures of unquestionable meteoritic origin. There are, however, three types of structure of greater magnitude which may represent the imprint of megameteorites on the surface of the Earth.

The first is represented by the Tunguska event: in this case, only, the trace of some form of observed mega-impact has been preserved: what it was that fell, however, no one knows. In contrast, the large cryptic craters and the crypto-explosion structures are imprints of events of very uncertain character, and in most cases there is no real certainty of extra-terrestrial involvement. In this section the two large craters which are associated with meteoritic material are included, because of the unresolved anomalies associated with them, and also included are the Henbury, Central Australian craters, which provide a link between the large impact explosion craters and small fragmentation craters.

The Tunguska Event

On 30 June 1908, at 0 hours, 17 minutes, and 11 seconds, universal time, a meteorite fell in the Tunguska River area of Central Siberia (60° 55′ N, 101° 57′ E). Optical, acoustical, and mechanical phenomena were observed over a large area surrounding the site of fall—a bright fireball moving from south-east to north-west, with a thick dust train, was followed by loud explosions heard up to 1,000 km away. Seismic and thermal wave effects were noted by observers over a wide area, and a shock wave was measured by observations at Irkutsk, Siberia, and Potsdam, Germany—both ground and explosive air waves being recorded. The latter were recorded, also, in England, and by other stations in Siberia. The energy involved in this event has been calculated as between 10^{21} and 10^{23} ergs (10^8–10^{10} MJ). There were effects on the upper atmosphere similar to those succeed-

247

Plate 69 Trees uprooted and felled by the Tungus event

ing major volcanic paroxysms of an explosive nature, and due to fine
dust, the nights were particularly bright over much of Europe and
Asia just after this event. The effects of this fall were not studied in
any detail till many years later. L. A. Kulik's expedition nineteen
years later found tall trees uprooted and radially felled in a zone
extending out 30–40 km from the focus, which was found to be a
marshy area 7–10 km in diameter, in permafrost terrain. Later ex-
peditions noted searing of trees up to 18 km from the focus.

The ground evidence confirms the explosive nature of the event,
indicating explosion on a scale not normally associated with meteorite falls. Further it does not go, and the lack of any meteorite recovery or crater is surprising. Visual evidence suggested a very low
angle of approach, and it is possible that a ballistic wave accompanying the meteorite was the real wrecking agent. Whipple and Astapovich attributed this event to a small comet, likely to have greater
geocentric velocity than is normal in the case of meteorites. The complete lack of material recovery is attributed to the nature of the mass

—ices of volatile components, sparse silicates dispersed in them, etc— and to a pulverisation effect in the course of atmospheric approach, producing a very fine dust from the mass.

The possibility that this was the impact of a body of 'antimatter' (more correctly Dirac-style matter) has been considered by several authorities. Whipple has considered the evidence against it to be overwhelming, but LaPaz considers that this possibility is worthy of serious consideration.

The Cryptic Craters

The arrival of very large meteorites, weighing thousands of tons, would, because of the fact that their cosmic velocity would be little dissipated by the atmospheric braking effect, result in explosion and vapourisation of the mass on impact. Certain large craters have been attributed to this process. They fall into two categories:

(a) Those with associations of meteoric iron.
(b) Those with no such association.

They are all of geologically youthful age—not older than the late Cainozoic (not more than ten million years). A list of these craters is given below (Table 20):

TABLE 20

Cryptic Craters

	Diameter, km	Reason for meteoritic attribution
Bosumtwi crater, Ghana	13	Coesite, glass of unknown origin
New Quebec crater, Canada	3·2	Structure, glass of unknown origin
Talemzane crater, Algeria	3·2	Structure
Lonar Lake, India	1·6	Structure, maskelynite, Shatter cones, Shock breccia, 'melted bombs'
Cañon Diablo crater, Arizona	1·6	Meteoric iron Shale balls Coesite, stishovite Impactite glass Rock flour Structure
Wolf Creek crater, Western	0·8	Structure

Australia		Shale balls
		Meteoric iron 6·5 km away
Pretoria Salt Pan, South Africa	0·8	Structure
Henbury craters, Central Australia	0·20 (largest)	Meteoric iron
		Shale balls
		Impactite
		Dispersion ellipse

Bosumtwi Crater

The largest of these is Bosumtwi crater, which has an association with breccias, clearly produced by some explosive process, and containing an acid glass referred to variously as dacite (an igneous rock type) or impactite. The crater is regarded by some authorities as of meteoritic origin, yet Junner, thirty years ago, produced an analysis which remains as convincing as ever. A geologist, he considered that the crater could be attributed to subsidence of cryptovolcanic origin. It is quite logical to expect that there would be some surface reflections of the process known as cauldron subsidence (which commonly takes place underneath large central volcanoes) that include the merest trace or no trace at all of surface volcanic eruption, and show cryptic brecciation of an explosive character. Junner regarded the difficulty of accounting for the enormous volume of rock material lost from the crater depression as critical, and it becomes more fatal to the meteoritic interpretation now we know that the crater is only between one and two million years old—geologically very recent—for this young age allows little chance of erosional removal of the material lost. This problem of accounting for the country rock lost in cratering becomes apparent in the case of many such structures (it is also a real one in the case of Wolf Creek crater, for example). In the case of Bosumtwi crater, Junner ruled out eviscerating explosion, basing his arguments on the structural deformation pattern and the complete lack of any blown-out material. His arguments have never been answered, popular as it is to regard this crater as of meteoritic origin.

New Quebec Crater

The New Quebec crater is a perfectly formed crater structure, and like Talemzane crater, and, significantly perhaps, Wolf Creek crater, it displays a radial pattern of topographically negative traces of fracture lines, nicking the walls. Currie has analysed the structure and considers it to be irreconcilable with extra-terrestrial origin. A fragment of glass of dacitic character recovered within it is unlike the country rock in composition, though Dence, who regards this crater as meteoritic, considers it to be impactite. Drilling has lately proved an association of igneous rock of apparently eruptive origin (Currie and Shafiqullah).

Telemzane Crater

Talemzane crater, Algeria, is twice the size as the famous Arizona crater. It shows structural similarities to both New Quebec and Wolf Creek craters. Karpoff suggested a meteoritic origin, but based this simply on comparison of the structure with other supposedly meteoritic craters. There are several other cryptic craters and circular structures in north-west Africa, including the Richat structures, Mauritania (which seem to reflect simple updoming of strata); and the Bosumtwi and Nebiewale craters in Ghana. There is, perhaps, a suggestive clustering of cryptic structures in this part of the world.

Lonar Lake Crater

Lonar Lake crater was recently studied by Lafond and Dietz, without any convincing evidence for meteoritic origin being discovered. It is perhaps significant that it is situated quite close to a carbonatite complex, one of the few known in India. Although the carbonatite is much older geologically than the crater, this relationship could still have some significance, either because of the observed fact that such explosive, alkaline volcanicity tends to recur along linear zones and at focal sites, and because of the possibility that these so-called craters (and here the Brent and Pretoria structures must be included with Lonar Lake) may be no more than erosional reflections of carbonatite diatremes. D. J. Milton (pers. comm, 14.3.72) has reported drilling by the Geological Survey of India, into the crater floor, revealing maskelynite basalt and some shatter coning. Strongly shocked debris and 'melted bombs' (sic) are reported in the ejecta.

Pretoria Salt Pan

The case here is very like that of Lonar Lake, except that carbonatite actually occurs as blocks within the structure (Milton actually talks of an exposure). The radiometric studies quoted by Milton suggest the age cf the carbonatite to be that of the nearby Premier Mine kimberlite and carbonatite (approaching 2,000 million

Plate 70 The Pretoria Salt Pan, South Africa; a cryptic crater-like depression on the Bushveld granite. The structure resembles the Wolf Creek Crater in scale and appearance, and in the polygonality of the crater rim

years, that is well back in the Precambrian), and the structure, if it is an explosion crater at all, whether impact or eruptive in origin, must be little older than Quaternary (3 million years) at the most. Wagner regarded it as a volcanic maar or explosion crater, but there is little evidence of volcanicity so late in time in this region, and it seems more likely to be an erosional reflection of an ancient diatreme, despite the fact that carbonatite pipes generally have a positive topographic expression in more youthful occurrences.

Cañon Diablo, Wolf Creek, and Henbury Craters

The last three craters in Table 20 are well established as of likely meteoritic origin. Wolf Creek crater is associated with shale balls of nickeliferous iron oxide (maghemite) discoloured by green, micaceous nickel minerals. Shale balls occur on the walls of the crater and outer slopes of the crater mound. They appear to represent bomb ejectamenta, having splash forms typical of volcanic bombs—pear-drop, triangle, banana, and disc, all with pronounced bread-crusting of the surface—and they occur welded into the laterite in what appears to be the position of fall. There is an unresolved anomaly both in the

252

nature and distribution of these shale balls. Octahedrite meteoric iron material has been recovered from a point 6·5 km from the crater. No impactite has been found, no coesite or stishovite, no shatter cones, and no rock flour. The deformation structure is anomalous in being highly asymmetrical, in spite of the symmetry of the crater, which displays very weak polygonality: one side is heavily deformed by concertina folds with broken hinges, whereas the other side has the form of a regular scarp of slightly upwarped stratified quartzites of probable Carboniferous age.

The structure must surely be congenetic with the Arizona crater, and is most probably meteoritic in origin, but in the author's view

Plate 71 Air photograph of Wolf Creek Crater. Seif dune and enclosing 'barchan dune' are evident, as well as the central gypsum area. The crater, 800 m in diameter, is markedly polygonal

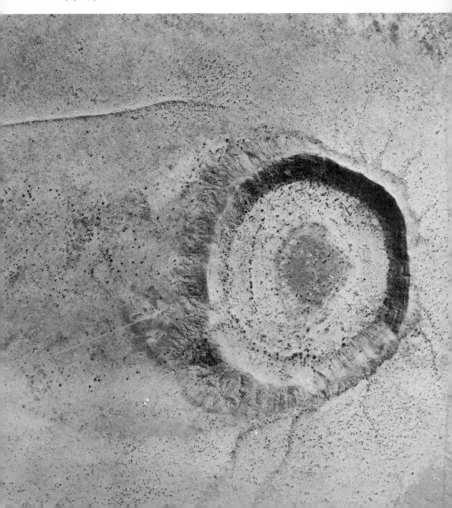

Plate 72 *Four iron shale balls from inside and just outside the rim of the Wolf Creek Crater. The form of volcanic bombs, a typical splash form, is clearly seen*

Plate 73 *Octahedrite iron fragment found 6·5 km from Wolf Creek Crater*

Plate 74 Cañon Diablo Crater or Meteor Crater, Arizona. Oblique view of the mile-diameter crater showing the rectangularity and crater-mound character

just an iota of doubt remains, particularly because both are situated close to kimberlite pipes—the Arizona occurrences have long been known, and kimberlite is now known at Mt Abbott, within 80 km of Wolf Creek. Drilling is called for in this structure: a major research programme should be mounted there, as in the case of Gosses Bluff and the Canadian structures. The structure is not dissimilar to the Pretoria Salt Pan in appearance and scale.

The Arizona crater is the archetypal impact explosion crater, having an association of octahedrite material profusely scattered around it, but not in it, and shale ball, impactite, rock flour, coesite, and stishovite associations. The latter are silica polymorphs produced experimentally at very high pressures. No well-formed shatter cones have been recognised.

Hager has advanced an alternative to the popular meteoritic explanation, publishing a map of the structure, showing it to be situated astride a linear belt of 'sinks and graben', and fissures from which lava emanated. He believed that the structure is controlled by this structural zone of internal origin and also by a fold pattern in the

255

Plate 75 Henbury Craters—oblique photograph

sedimentary rocks. Some of the anomalies recognised by Hager seem not to have been explained.

There seems to be little doubt that the iron is meteoric; though cosmic-ray induced isotope data seem to be lacking, it otherwise appears to display meteoric characteristics. This seems overwhelmingly to favour meteoritic origin for the crater, unless there are two superimposed occurrences. The crater is certainly very young, much younger than any so far described, and this could account for the lack in other craters of some of the manifestations associated with this one, the original crater wall having elsewhere been cut-back by erosion and the intense modification effects having been lost. An age of 6,000 years has been indicated. The impacting mass would have weighed 30,000 tons, it is estimated, vapourising itself and partly melting the sandstone rock nearby because of its unabated cosmic velocity. Many small glass spherules found for miles around the crater are supposed to be scatterings produced by the melting of the rock.

The Henbury craters are a group of large and small structures: the largest, 200 metres across, is probably an impact explosion crater, whereas the smallest are probably fragmentation craters: in this way one can account for the association of metal chunks, twisted small iron masses, shale balls, and impactite glass. The craters form a pattern suggestive of a dispersion ellipse. The iron is of octahedrite type.

256

The Arizona and Wolf Creek structures are the only two really large terrestrial crater structures for which a strong case for meteoritic origin can be said to have been made. The meagre representation of such large craters renders the attribution of other large crypto-explosion structures to extra-terrestrial agencies the more dubious. Comparisons are constantly being drawn, but even the type examples are themselves to some extent ambiguous, and the question must be asked whether comparisons are not being drawn, over and over again, between structures of obscure origin and other such structures!

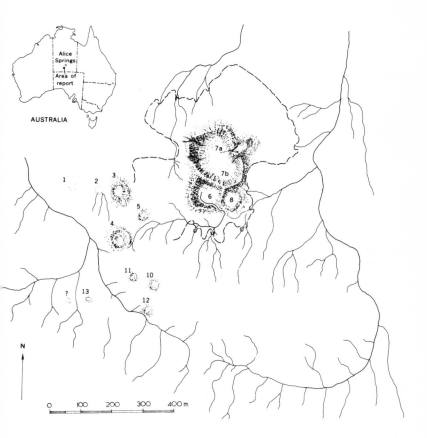

Fig 20 The Henbury Craters, showing the typical dispersion ellipse pattern (after D. J. Milton and F. C. Michel)

Crypto-explosion Structures, Possible Astroblemes

Hey (1966, p 538) presents a formidable list of structures for which a meteoritic origin has been suggested at one time or another, but the structures included are, in almost all cases, of doubtful meteoritic origin.

Possible Criteria for Recognition of Extra-terrestrial Agency

Criteria used to support meteoritic attributions are:

Shatter cones.
High-pressure silica polymorphs.
Shock micro-deformations of quartz, etc (lamellations).
Impactite glasses.
Traces of nickel.
Nickel-iron associations.
Structure.

In Table 21, the important attributions are listed, alongside the grounds for such attribution and comments from the present author.

Shatter cones have never been directly linked with impact or impact-explosion, and their status as an indicator of meteorite involvement seems to have been questioned as a result of finds at Vredefort and Sudbury, vast structures extremely difficult to equate with extraterrestrial agencies. Confocality of shatter cones to a point of impact has been suggested for Vredefort and Gosses Bluff, but shatter cones at Steinheim show three orientations in one hand specimen, and though the confocality at Vredefort has been elegantly demonstrated by Manton, unrolling the subsequent deformation, Nicolaysen has advanced quite another explanation in terms of endogenous diapirism.

Coesite and stishovite, the high-pressure silica polymorphs, remain only valid indicators of extreme, possibly explosive pressures; there is no real evidence that they are restricted to extra-terrestrial deformations. Coesite can exist metastably outside its theoretical and experimental range of pressure and temperature, and supposed restrictions on pressure involvement in eruptivity based on the strength of the retaining crust are suspect. Traces of nickel geochemically determined prove nothing—the recent nickel exploration boom has established the ubiquity of minor nickel geochemical anomalies in rocks.

258

Structures Possibly of Meteoritic Origin

Structure	Nature	Dimensions, diameter (km)	Basis for meteoritic attribution	Remarks
A Large ring structures, etc. Vredefort Dome, South Africa	Cylinder of Archaean granite pushed up with overturning of collar	56	Shatter cones, overturned collar	Impact origin favoured by Dietz, Hargraves, and Manton: endogenous origin favoured by Bucher, McCall, and Nicolaysen.
Bushveld Complex, South Africa	Vast series of associated eruptive complexes, of ultrabasic, basic, and acid igneous rocks: annular patterns	320 × 160	Relation to the Vredefort Dome	Impact origin suggested by Dietz, but not seriously entertained by most geologists. Both these structures have some genetic connection to the Vredefort Dome, and it is difficult to see how any one could be of extraterrestrial origin without the whole group being so.
Great Dyke, Rhodesia	Vast linear complex of similar igneous rocks	1130 × 11·3 (max)	Relation to the Vredefort Dome	
Sudbury Irruptive, Canada	An elongated ring of confocally dipping layered igneous rocks, similar in some ways to the Bushveld Complex	65 × 24	Shatter cones, shock lamellation of quartz, etc, supposed impactites	Though geologists employed by the International Nickel corporation do entertain meteoritic triggering seriously for the vast magmatic emplacement, no one now claims that the nickel sulphide ores are of meteoritic origin (and this seems to have been the original reason for suggesting impact involvement)
B Crypto-explosion structures Mississippi Belt	These are complex structures, involving arcuate and linear, commonly radial faulting, internal depressions, and central uplifts. Extreme deformation, brecciation, and shatter coning are characteristic	Up to several kilometres diameter	Shatter cones, structural arguments shock lamellation of quartz: coesite (a high-pressure silica polymorph) present in the case of Kentland	These structures are of widely diverse ages: they have an association with lead ores: they form a distinct zone following in a broad manner the Appalachian orogenic trend: igneous rocks of normal type may be present: they show evidence of a systematic tectonic control on their situation: they follow a belt characterised by several diatremes, eruptive complexes characterised by gas drilling (for example, Magnet Cove carbonatite complex)
A group of structures recorded by Bucher as cryptovolcanic, together with Upheaval Dome, Utah, Riesskessel, and Steinhem Basin. Many more now known, all in a broad belt running parallel to the Appalachian Mts, inland from them				
Kentland Structure, Serpent Mound, Jepthas Knob Crooked Creek Structure, Furnace Creek Structure, Decaturville Structure, Rose Dome, Hazel Green Creek Structure, Weaubleau Structure, Hicks Dome, Wells Creek Basin, etc, etc				

TABLE 21—(*contd.*)

Structure	Nature	Dimensions, diameter (km)	Basis for meteoritic attribution	Remarks
Ontario 'Craters', Canada Brent Crater, Holleford Crater, Lake Wanapitei, Franktown Crater	Several ancient infilled crateroid depressions, associated with breccias	Diameters up to a few kilometres	Structural arguments; coesite present, in at least one case. Shock lamellation of quartz, etc.: glassy melt breccias, melt rocks	Their varied ages, situation on a northwards extension of the Mississippi Belt, and the coincidence of their areal distribution with carbonatite and diamond distribution suggest endogenous origin. The Brent Crater has revealed rocks of carbonatite associations on drilling and a fenite contact zone, characteristic of carbonatite complexes
Other Canadian structures Deep Bay, West Hawk Lake, Manacouagan Lake, Clearwater Lakes, Lake Malbaie, Carswell Lake, Mecatina Structure, Natapoka Arc, Hudsons Bay, Gulf of St Lawrence Arc Greenland structure, Agto, W. Greenland	A whole list of meteoritic attributions has stemmed from Beals, Dence, Innes, *et al.* These are mostly circular structures of great age, some crudely crateroid	These are rather larger than the Ontario Craters, diameters measuring tens and even hundreds of kilometres	Structural arguments; supposed impactites; shock lamellation of quartz, etc.: glassy melt breccias, melt rocks, maskelynite	There is really little real evidence for extraterrestrial origin other than indirect arguments based on comparisons with other doubtful structures. Shock lamellations are probably the strongest evidence. The distribution is suggestive of a splaying out of the northern end of the Mississippi-Ontario belts, and they are strongly clustered in one part of Canada. Clearwater Lakes and Manacouagan have what appears to be a dacite lava association (but Dence regards it as a strange form of impactite). Igneous rocks are proved by drilling in depth close to several of these structures. Currie regards these as resurgent calderas. Bostock favours endogenous origin also

TABLE 21—(contd.)

Structure	Nature	Dimensions, diameter (km)	Basis for meteoritic attribution	Remarks
South German Structures Rieskessel, Steinheim Basin	The first is a crateroid structure associated with complex outthrusting deformations: the second is a structure with an uplifted central area, and fault patterns resembling the Mississippi structures	The Ries is 22·5 km in diameter, the Steinheim basin much smaller	Suevite glass (supposed impactite): shatter cones; coesite: traces of nickel-iron in the suevite: structural arguments: shock lamellation of quartz, etc	Bucher rejected impact involvement on grounds that the stratigraphy indicates a multistage process of formation (Ries): that the Ries is situated at the confluence of two major geotectonic lineaments: that there is a geothermal focus (Steinheim): that geophysics indicates a dense igneous core in depth: and that the two structures line up with the 'vulkanembryonen'
Scandinavian Occurrences Lake Hummeln, Sweden Lake Mien, Sweden, Lake Lappajarvi, Finland	Obscure sites of cryptic deformation	A few kilometres across	Structural arguments, coesite, shock lamellation of quartz, etc	
Australian Occurrences Gosses Bluff, Central Australia	An immense circular uplift, accompanied by much deformation, brecciation, shatter coning	Several kilometres diameter	Structural arguments (overturned collar): shatter cones: shock lamellation of quartz, etc	The structure penetrates at least 10 km in depth: geophysical evidence suggests that there are undisturbed strata continuous beneath it, and this has been considered an argument for impact origin: however, this will be the case whatever its origin: geological structures just do not continue indefinitely in depth: the structure appears to be situated at the intersection of crossing folds
Liverpool Crater Northern Territory	A crater-like ring of sandstone breccia	1·6 kilometres	Structural arguments, some shock deformation of quartz; possible impactite glass	
Strangways structure, Northern Territory		≃16 kilometres	Structural arguments; shatter cleavage: possible shatter cones	

261

Plate 76 Shatter cone, from the Kentland crypto-explosion structure, Illinois

Impactites are difficult to identify as such, unequivocally, unless there is a close association with meteoric material. Glasses such as the suevite from the Rieskessel and the Bosumtwi or Clearwater Lakes dacite could well be unusual types of silicic volcanic glass, either of igneous origin or due to secondary melting of country rocks in depth as a result of primary igneous agencies. Symmetry of structures, of many forms, has been cited as evidence for impact involvement, but volcanic structures display every type of symmetry and this criterion is certainly invalid.

Shock effects microscopically evident—the lamellation of quartz grains, etc—are commonly found in cryptic structures on Earth and in lunar surface rocks. As with the silica polymorphs, however, no convincing case has been made for the exclusive occurrence of such effects in impact environments, and far more work on terrestrial deformation strains of other sorts must be done before such a case can be accepted.

The structural arguments are based mainly on comparisons between one enigmatic structure and another, and, as previously noted, there is no type example that can be unequivocally used as a model, except possibly the Arizona and Wolf Creek, and largest Henbury,

262

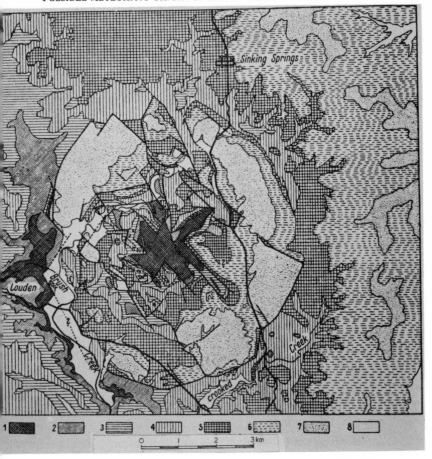

Fig 21 The Serpent Mound, Ohio, crypto-explosion structure. The formations are differentiated by ornament contrasts on this geological map: they are various types of sedimentary formation; faults are delineated by heavy lines (after W. H. Bucher)

craters. Impact explosion, diatreme activity exemplified by diamond (kimberlite) pipes, carbonatites, and lamproites, crypto-volcanic explosions, and phreatic explosions involving interaction between magma and groundwater, are all likely to produce similar deformation strains, and there would appear to be no criteria that unambiguously separate them, despite the important studies of artificial nuclear explosions. Brecciation effects and melting effects are likely

263

to be similar. Rapid decreases in brecciation downwards, and surprisingly deep penetration downwards of the breccia zone, as at Gosses Bluff, have both been recognised, and a finite bottom to the breccia zone is the expectation of both extra-terrestrial and endogenous structures, for in neither case can the mechanical deformation go down indefinitely.

The association of nickel-iron, or other meteoritic material, remains the best of all criteria but, even then, the possibility of overprinting of structures or occurrences of diverse origin remains real—no less than ten different meteorite events are indicated in an area less than twenty miles long in the arid Nullarbor plain of Western Australia! And the possibility of material simulating meteoric iron being ejected from deep within our own planet cannot be entirely overlooked.

Geographical Groupings

The geographical grouping of crypto-explosion structures has been used by Dence in the case of the North American occurrences (numbering more than sixty) to deduce that an impact explosion crater fifteen kilometres or more in diameter is formed every two million years on the land surfaces of the earth! But the same grouping may be used to question the validity of the whole range of attributions. For, there is not only a heavy concentration in North America, but there is also, within that continent, a strong concentration along a linear zone following the trend of the Appalachians, as if geotectonically controlled and splaying out northwards into Canada, where there is coincidence with carbonatite and diamond occurrences—both indicative of endogenous explosive eruptivity. The remainder of North America has revealed no greater number of structures than the rest of the world, so it is not just a case of greater scientific activity in the search for such structures in North America! In the linear belt mentioned above the craters are of widely different ages and many are now being revealed as in all probability of endogenous origin: there is a strong resemblance between this repetition of similar central eruptivity (?) patterns along a linear zone and the pattern of development of anorogenic linear zones such as the Rift Valley Zone of Eastern Africa.

Other groups are the South German structures, the North West African, and Central Australian structures; a group in Scandinavia is also becoming defined. The Rieskessel and Steinheim Basin are lined

Plate 77 The Gosses Bluff crypto-explosion structure, Central Australia

up with the nearby *vulkanembryonnen*—a cluster of small diatremes of accepted endogenous origin—and there is geophysical evidence of a dense igneous body beneath the Rieskessel. The stratigraphy suggests, according to Bucher, a multistage formation process, not a simultaneous one, such as impact explosion: he believed that the Ries formed in successive stages, appreciably spaced in time. He also considered the situation of the Ries exactly on the junction of two prominent geotectonic lines to justify his remark 'if it was a meteorite it was a highly educated one!'

The Gosses Bluff structure in Australia, one of the newest discovered and finest of all the cryptic structures, is associated with shatter cones, bentonite, and zeolite mineralisation. The latter two mineralogical associations are commonly associated with volcanic processes. It penetrates to at least 10 kilometres!

These structures are discussed again in Chapter 21 under tektites, for some of these structures appear to have some genetic connection to tektites, though the exact nature of this genetic connection remains obscure.

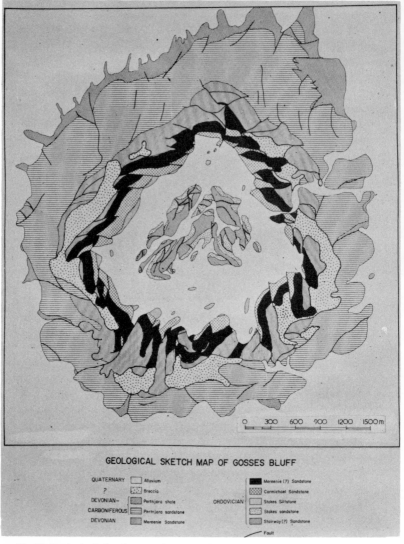

GEOLOGICAL SKETCH MAP OF GOSSES BLUFF

QUATERNARY		Alluvium			Mereenie (?) Sandstone
?		Breccia			Carmichael Sandstone
DEVONIAN–		Pertnjara shale	ORDOVICIAN		Stokes Siltstone
CARBONIFEROUS		Pertnjara sandstone			Stokes sandstone
DEVONIAN		Mereenie Sandstone			Stairway (?) Sandstone
					Fault

Fig 22 Geological sketch-map of Gosses Bluff (from Records, Bureau of Mineral Resources, Canberra)

Summary of Conclusions

The following conclusions are drawn by the author:

(a) Impact explosion and cryptovolcanic explosion are processes that will produce similar effects, and thus are not likely to be readily distinguishable on any criteria except association of meteoritic material or familiar igneous rock types.

(b) Even then the remote possibilities that eruptive or volcanic rocks use meteoritic structures as access channels to the surface (as they apparently utilise non-eruptive salt diapir structures) or cryptovolcanic explosion brings up meteorite-like material to the surface cannot be entirely ignored.

(c) Impact explosion and crypto-explosion are unlikely to be distinguishable on the criteria of energy involvement.

(d) High-pressure silica polymorphs, shock lamellation effects, shatter cones, and 'impactite' associations are all ambiguous.

(e) Geochemical traces of nickel do not afford a valid criterion.

(f) Symmetry does not provide a valid criterion.

(g) Structural analogies are ambiguous, especially because of the lack of a sound foundation for comparison.

(h) Geographical evidence of clustering and alignment, and evidence of varied ages within individual geographic clusterings, are against meteoritic arguments.

(i) The record of megameteorite impacts on the Earth's surface is surprisingly meagre, as is the record of fossil meteorites in the rocks.

(j) The Tunguska event seems to have been something different from a simple impact of a mass with high mass and geocentric velocity. It may have been friable like some enstatite achondrites and so readily pulverised: it may have been a cometary impact, of a mass of ices, and dust: special conditions of extreme low angle of incidence might even have produced the catastrophic effects.

(k) Wolf Creek crater, Australia, and Cañon Diablo crater, Arizona, are the only two well-established examples of mega-impacts besides the Henbury craters: and slight anomalies and uncertainties are evident in even these two cases.

(l) The lack of mega-impact scars on Earth suggests that the number of meteoroids orbiting in space, with orbits of Mars-passing geometry bringing them into possible contact with the Earth's orbit, may have been exaggerated.

267

(m) Chemical explosions involving prolific releases of gases such as hydrogen may be involved in endogenous nupto-explosion structures, this accounting for the high shock pressures.

21: Tektites—Nature and Occurrence

The term 'tektite' was coined by the geologist, F. E. Suess, in 1900; however, the *moldavites*, from what is now Czechoslovakia, were known and described late in the eighteenth century, and the australites were described and discussed about 1840 by Charles Darwin, who regarded them as a special form of 'obsidianite volcanic bomb'. The possibility of extra terrestrial origin does not seem to have been considered until early in the present century.

Tektites are small glass objects, consisting of an acid glass—that is of the composition of a silicic igneous rock and not pure silica, though they do contain the pure silica glass, *lechatelierite*, as minute, commonly shred-like inclusions. In a very few tektites from the Philippine Islands small spheroids of nickel-iron (kamacite), nickel-iron sulphide (troilite), and nickel-iron phosphide (schreibersite), all mineral components characteristic of meteorites, are found. This and the fact that the Australasian tektites show ablation effects, to varying degrees, are the only features suggesting a meteoritic origin. Tektites, however, do not occur all over the Earth like meteorites, but are limited to geographical areas known as 'strewn fields'. These are all situated at relatively low terrestrial latitudes. The four groups known, of different ages of arrival, are listed in Table 22.

TABLE 22

Tektites

Group	Locality	Type/Name	Age of arrival
4 Australasian	Southern Australia	Australites	Middle (Late) Pleistocene
	South-East Asia		
	Extreme South China		
	Indochina	Indochinites	
	Malaysia	Malaysianites	
	Philippine Is	Philippinites	
		Rizalites	
	Indonesia	Billitonites	
		Javanites	

269

TABLE 22 (*contd.*)

Tektites

3 West African	Ivory Coast	Ivory Coast Tektites	Lower Pleistocene
2 European	Czechoslovakia	Moldavites	Miocene
1 North American	United States		Oligocene
	Texas	Bediasites	
	Georgia	Georgiaites	
	*Martha's Vineyard		

* One solitary occurrence.

Tektites do not arrive on Earth continuously, like meteorites (*sensu stricto*): they arrived, whatever their origin and provenance, in at least four separate arrival events, all during the last geological era—the Cenozoic, which comprises the Tertiary and Quaternary, and commenced about 60 million years ago—even then they did not arrive until well into the Tertiary, about 34 million years ago. There is no certainty that they are of extra-terrestrial, origin, that they do not come from our own planet—indeed there is some evidence to suggest that this was their origin, yet some at least of the tektites have suffered ablation while coming in through the Earth's atmosphere. Surprisingly, perhaps, there is less reason to connect them with true meteorites than with the Earth, the Moon, or cometary sources.

There is no trace of tektites at all in the 3,500 million years of the Geological Record, preserved in the terrestrial rocks, and their limitation to the last instant of geological time, as it were—the last 34 million years—is, perhaps, significant. Fossil molluscs and coprolites (animal excreta) are copiously preserved in rocks, and one may well ask, 'why not tektites in the ancient rocks, if they did arrive on Earth in a uniform manner throughout the long history of the Earth?' The lack of fossil record of true meteorites is puzzling, but can be explained by the lack of very diagnostic shapes and the chemical nature of meteorites, which allows rapid decay: the tektites have, at least in the case of the *australites*, forms as diagnostic as many fossils, and their chemistry would not lead to very rapid decay: why are the shapes not found fossilised? This lack may be due to their manner of etching—all the older tektite groups show very irregular etched forms —but it may indicate that tektites were relatively late arrivals on the

Plate 78 Upper left, bediasite, Texas; upper centre, upper right, lower right, and middle right, moldavites, Czechoslovakia; middle left and lower left, Darwin glass, Tasmania; middle centre and lower centre, billitonites, Indonesia

geological scene, and whatever their origin, cannot be extrapolated backwards in time as an intermittent, though now obscured feature of the geological history.

While the problem of the ultimate provenance of tektites remains

271

unresolved we cannot regard them as meteorites—for meteorites are by definition of extra-terrestrial material. At present there is a conflict of evidence concerning the origin of tektites, a conflict that is only complicated by the introduction of refined instrumental methods of research over the last decade. They appear to be *sui generis*, one of the unexplained Natural Science phenomena. It is only for convenience and by custom that they are included in the discipline of 'meteoritics'.

Tektites cannot be divorced from certain enigmatic occurrences of cryptic glasses at Mt Darwin in Tasmania and in the Libyan Desert: nor from associations of other cryptic glasses with very large crypto-explosion structures and craters, especially the Rieskessel/suevite association in South Germany and the Bosumtwi crater/cryptic glass association in Ghana. Similarly, the possibility that they are related to impactite glasses such as are found at Cañon Diablo crater, Arizona; the Wabar craters in Arabia, the El Aouelloul crater in West Africa and the Henbury craters in Central Australia, and are attributed to the melting of the country rock by the agency of large meteorite impacts, cannot be entirely discounted.

Tektites have three basic forms according to Dr Virgil Barnes:

(a) Irregular forms—exemplified by his Muong Nong recoveries from Laos, Indochina, where such irregular tektites were first discovered.
(b) Primary forms of rotation—spheres, dumbells, and apioids.
(c) Secondary forms derived from shapes of type (b) by ablation, melting off, and shedding material during atmospheric entry on account of friction.

The fact that all these three types exist and have to be explained by any satisfactory theory for tektite genesis is one that is commonly overlooked. Australites show the most perfect ablation forms (type c). Some javanites show evidence of ablation, though with no ring-wave patterns preserved, but other non-Australian tektites do not show distinct ablation forms and it appears that the primary forms, which have a similar range in all tektite groups, have not suffered ablation loss to any measureable extent.

The Primary Phase

Tektites of types (b) and (c) have a range of primary shapes evident, which are attributed by Dr George Baker to various speeds of rotation while the molten drop cooled to form the glass body. The primary forms are spherical or elongated, and on these forms ablation super-imposed flange and ring-wave patterns. The equatorial flange tends, however, to break off and the common recovery, in the case of type (c) tektites (australites), is an ablated sphere, dumbell or apioid, showing an annular equatorial scar marked by shallow, parallel scallopings: this form is known as the *core*. Spherical primary shapes will be ablated to a *flanged button* ideally, but the *button core* is the most common australite recovery: flanges are occasionally recovered as separate tektites.

The primary forms of tektites are:

Spheres.
Prolate spheroids.
Oblate spheroids.
Dumbells.
Apioids of revolution.
Canoes.

The primary body consisted of solid tektite glass with minute en-closed bubbles: in rare cases the bubbles are quite large, and a few tektites consist of a thin glass shell enclosing a bubble which forms the greater part of the volume of the tektite.

In suggesting a rotational control on the development of the primary forms Dr Baker recognises:

Spheres——Oblate Spheroids——
                             Prolate Spheroids——Dumbells Apioids

No rotation——increasing rotation—— . . . *——* . . .

The dumbell is supposed to form when there is centrifuging of the molten glass up and down the axis of rotation, whereas the apioid is produced when the waist of the dumbell becomes so narrow that separation into two bodies occurs. All sections normal to the axis of rotation were essentialy circular prior to ablation modification. *Schlieren* and bubble pits are revealed throughout the entire body of a tektite: the former, streaky layers which disregard the attitude of ablation-produced surfaces, and the bubbles, are attributed to entrap-ment of volatiles in the gas phase during the primary solidification. Flanges produced during ablation do, however, show schlieren which conform to the shape of the flange and are clearly features of solidifi-cation after secondary melting during ablation.

273

Plate 79 Some dumbell-shaped tektites. (a) An indochinite from Thailand; (b) an australite from Cuballing, Western Australia; (c) the same, end-on view; (d) an australite (British Museum No 50918)

The primary forms are less well defined in the case of non-Australian tektites, especially the older African, European, and American groups, which show highly etched glass surfaces. Whether they had less perfect primary shapes in the first place or, alternatively, long sojourn in soil and rock and overland transport by streams has inflicted chemical and mechanical destruction on their surfaces, is not certain.

The Ablation Phase

Types (a) and (b) of Barnes do not display the effects of atmospheric ablation. The flanged button type of australite, found perfectly preserved at Port Campbell, Victoria, has been taken as a model in simulation experiments by Dr Dean Chapman, in America, and he has, with his associates, reproduced this form using gelatine spheres and tektite glass in oxy-acetylene arc ablation experiments, so that there can be no doubt that these are ablation forms. It is indisputable that ablation formed these objects, and it seems reasonable to suppose that it was atmospheric ablation (though one may question, in passing, whether ablation in a fierce blast of gas erupted from the surface of the Earth could not produce such shapes): it is generally accepted that atmospheric friction acting on initially cold, non-rotating, rigid glass bodies, of one of the primary forms, produced these ablation forms during atmospheric entry at supersonic velocities. If these shapes are only compatible with such atmospheric ablation, as Chapman believes, then it seems certain that the australites, and at least some Indonesian tektites, must have been hurled out through the Earth's atmosphere and returned through it, if tektites are of terrestrial origin.

During ablation the melt glass was forced from the stagnation point, situated in the anterior polar position during stably oriented flight, outwards and backwards past the equatorial zone, to form a backwards directed flange, in which it was re-frozen, without boiling and with little bubbling. Round, oval, elliptical, and dumbell-shaped objects all developed such flanges, with resultant reduction in the anterior surface from the primary circular section to give a reduced lenticular sectional outline, the original curvature being preserved on the posterior surface within the flange, while the front surface shows a reduced curvature. No other strewn field has yielded flanged tektites. Even in the case of the rare south-east Asian tektites, the crude form of which shows some sign of ablation, the loss seems to be

Plate 80 Photograph of an artificial wind-tunnel ablation product of tektite glass compared with a real australite (below)

rather slight for the cross section remains close to the primary circular cross section.

The Terrestrial Surface Phase

Tektites, once arrived on the surface of the Earth, were subject to erosive agencies, redistribution by stream and sheet flood action; and also to marine agencies. The earliest arrived tektite group, those of North America, will have had much longer to suffer erosive effects and secondary redistribution. It is, in fact, unlikely that the term strewn field has much validity in the case of such early tektites: even the situation of the later moldavites may not correspond very closely with the actual strewn field. The Pleistocene groups are probably found in distributions much more closely coincident with the actual strewn field, and this applies particularly to the australite field of recovery, much of which is very flat terrain not subject to the action

276

of powerful streams: redistribution in such terrain would probably be mainly a localised effect, moving the tektites into the intracontinental lake basins, where in fact they are found in profusion—more than 30,000 terrestrially etched tektites have been recovered from Lake Yindarlgooda in Western Australia, and some 10,000 from a lake near Israelite Bay in the same State. Few tektites, even australites, are found in the site of fall, though dune recoveries, and the very fresh tektites from Port Campbell (which have clearly suffered negligible transport) probably represent recoveries more or less in situ. Strewn fields, then, are, strictly speaking, approximations, and in the case of the older groups, the approximation between field of recovery and site of fall may not be very close: in such cases the actual site of recovery is where sediments of the right age, or later sediments containing reworked material are exposed at the surface. South-east Asian tektites have been recovered from the sea floor, and 'microtektites', supposed to be related to the Australasian group, have been recovered from the floor of the Indian Ocean (however, their identification as tektites remains uncertain).

The controversy regarding the solution etching of tektites remains not entirely resolved. Vermiform patterns of etching on the surface of tektites are familiar to any worker in this field of research: Barnes and Baker have long accepted these patterns as being of terrestrial origin, due to weathering: lately, however, Dr John O'Keefe has made a case for 'extra-terrestrial' etching during ablation, before arrival. Accepting the three basic primary forms of Barnes (a, b, and c), and noting that Suess, in 1900, favoured aerodynamic sculpturing, he makes the following points:

(a) Artificial attack by hydrofluoric acid follows the lines of internal schlieren, while natural sculpturing for the most part ignores these controls.

(b) The patterns seem to have a systematic relationship to the shape of the tektite.

(c) No silica residue is left in the form of a skeletal microstructure, but such a microstructure does result, as a residue, when volcanic glasses are corroded (to this one might add that the vermiform patterns of tektite etching are not commonly seen on the surface of volcanic obsidian masses exposed to terrestrial weathering agencies).

(d) The Texas occurrences include fresh and sculptured flangeless

277

forms recovered from the same site, suggesting that the control on etching is not ground chemistry.

(e) Tektite-like objects recovered from the sea bottom reveal similar sculpturing to land surface recoveries.

(f) Experiments on aerodynamic flow under certain conditions has been found to produce cavitation in glasses. However, the actual, less dense tektite glasses did not cavitate in this way under experimental conditions.

(g) Nininger and Huss describe a tektite from Indochina on which the grooved and pitted etched surface is clearly seen to be of prior age to plastic deformation of the elongated tektite, deformation which has arched it up longitudinally with the partial opening up of a crack at the point of flexure.

O'Keefe's arguments do leave the possibility open that some of the surface etching of tektites may be due to sculpturing in ablation flight rather than ground corrosion, but while this may conceivably be true of the neat vermiform groovings, it is certain that some etching is terrestrial, and due to ground corrosion. The tektites recovered from the salt lake of Yindarlgooda in Western Australia are all very etched, in a very irregular fashion involving minute pitting, and this certainly reflects the fact that they have been washed into the lake along the drainage system—they are recovered mostly from small fans of small quartz and ironstone pebbles where the streams debouch into the salt lake. Against O'Keefe's arguments is the fact that the tektite groups yield deeply etched and less etched specimens, and the oldest group, the Bediasites, include individuals showing very deeply incised crevices on their surfaces: the Moldavites, half their age, include individuals showing not so deep crevices, while the Billitonites show such crevices only starting to form. The state of the most etched specimens does seem to show a progressively increased development of etching with terrestrial age. The contrast between the Billitonites and the australites may be in part due to climate—the high, tropical rainfall and tropical soil chemistry may have something to do with the etched state of the south-east Asian tektites compared with the apparently contemporary australites. This is an irregular pitting and may not be the same process as produced the very neat groove patterns: and the possibility still remains that the neat grooving is not a ground corrosion effect, even if the irregular pitting, leading to surfaces composed of anastomosing cracks separating raised plateaux and peaks, is due to ground corrosion. If sculpturing is in part ablationary, we must be careful in invoking evidence of sculpturing as evidence against reported observations of tektite arrivals. Though the evidence against such observations is in most cases convincing, and there is little doubt that tektites do not arrive intermittently like meteorites, thus gradually and continuously increasing the store of material on Earth, some reserve must be maintained unless the nature of the etching is clearly of terrestrial type: etching in itself cannot otherwise be taken as absolute proof of long sojourn in the soil, etc, especially if it is restricted to the neat grooving type.

The Age of Tektites

There are three possible ages to be determined (according to Dr E. Anders, on whose work this section is largely based):

(a) The age of solidification of the primary object.
(b) The age of ablation on atmospheric entry (age of fall to Earth?).
(c) The age of the enclosing sediments, or the formation upon which the tektite rests at the time of recovery.

(a) *Solidification Ages*

The solidification ages of tektites were studied by Schnetzler and Pinson in America, using $^{87}Rb/^{87}Sr$ decay. The method used is the generally accepted modification of the original method involving the simple ratio above, being what is known as the isochron method, involving also the $^{87}Sr/^{86}Sr$ isotopic ratio (it is the same method as has been used for determinations of solidification ages of true meteorites, Chapter 18). Using this method, the ages of indochinites, philippinites, javanites, australites, moldavites, bediasites, and the solitary Martha's Vineyard tektite have been determined. There is a very small range of initial strontium ratios revealed in comparison with the wider range displayed by terrestrial igneous rocks, the values being between 0·7121 and 0·7223. The $^{87}Rb/^{87}Sr$ ratio is found to be very similar for specimens taken from the same locality, but there are slight differences between localities. For all the groups studied, ages close to 400 million years (equivalent to the Carboniferous period of the Palaeozoic era, geologically speaking) were obtained. This is a really surprising figure for primary melt solidification and difficult to

279

reconcile with any theory of origin so far advanced. It is certainly difficult to reconcile with any theory of lunar provenance, unless the Moon was differentiated within the last 400 million years or so (which seems unlikely)—the tektites could conceivably represent impactite, or ejectamenta from lunar volcanoes since this date, if this were the case. They certainly could not be derived from acid (granitic) material formed at the generally accepted solidification age for the Earth and other bodies of the solar system (Chapter 18); if this were so the initial strontium-86/87 ratio would be about 0·890. A source in chondritic material is also excluded—the initial strontium ratio would in that case be about 0·750, still much higher than the actual value. Differentiation from a basaltic parent melt is considered theoretically possible by Schnetzler and Pinson, but they still consider the values obtained are radically different from those expected on the basis of any theory of ejection from the lunar surface. They do not, however, go further than this, and, so far, these results have not been reconciled with any theory.

(b) *Ablation Ages*

The ablation age is derived by potassium-argon methods, based on the decay of ^{40}K to ^{40}Ar: it can also be derived by a method based on fission track counts, the measurement microscopically of the number of tracks left on a glass surface by fission products and obtaining a ratio against the number produced by artificial exposure of the specimen to a known radiogenic source for a known period of time. It must be noted that the latter method relies on the same decay constant as the $^{40}K/^{40}Ar$ method, and any inaccuracy in determination of this constant will be shared by the two methods. Both methods give the age of solidification of the glass during ablation, not the primary solidification—it has been found that the very heart of a flanged button tektite yields the same potassium–argon value as the flange, which was solidified and only became a close system for the noble gas daughter element during ablation. This method, therefore, gives the *age of arrival on Earth*. Unlike the rubidium/strontium method, this method cannot be used to probe back into the pre-ablation history of the tektite: the whole tektite became a closed system for ^{40}Ar only on ablation heating.

The ages obtained by this method are shown in Table 23.

TABLE 23

Tektite Ablation Ages: All Ages in Millions of Years

Group 1	South-east Asian tektites	0·61†	K/Ar	
	Australites	0·61†	K/Ar	
	Kalgoorlie	0·34 (±0·01)	F/T*	
	Kalgoorlie	0·66 (±0·01)	F/T*	
	Port Campbell	0·11 (±0·01)	F/T*	
	Port Campbell	0·69 (±0·05)	F/T	
	Port Campbell	0·69 (±0·07)	F/T	
	Charlotte Waters	0·64 (±0·16)		
		0·65 (±0·01)	F/T	Mt. Darwin, Tasmania (cryptic glass).
Group 2	Ivory Coast tektites	1·3†	K/Ar	
		1·3	K/Ar	Bosumtwi Crater, Ghana (cryptic glass).
Group 2	Moldavites	14·7 (±0·7)	K/Ar	
		14·8 (±0·7)	F/T	Rieskessel, South Germany ('suevite'—cryptic glass).
Group 1	Bediasites	34·9†	K/Ar	
		33·4†	K/Ar	
		34·2 (±0·8)	F/T	
		34·5 (±1·5)	F/T	
	Georgiaites	33·7†	K/A	
		34·5 (±1·4)	F/T	
	Martha's Vineyard	33·9	K/A	
		33·7 (±2·2)	F/T	
				Clearwater Lakes, Quebec, Canada.
		33·5 (±4·5)	F/T	(Cryptic glass).
		33·8 (±2·5)	F/T	Libyan Desert (cryptic glass).

K/Ar Potassium/argon age.
F/T Fission track age.
† Average value.
* Fleischer and Price, who made these determinations, conclude that there were *either* four different ablation events applicable to the australites, *or* the track numbers have been altered during the period on Earth so as to give new and quite spurious apparent ages in three localities. They favour the second alternative: yet it is not without possible significance that low values have been obtained on some of the most perfect australites from Port Campbell, which Baker and Gill (p 282) have attributed to much more recent falls than other australites, on the basis of other lines of evidence. In noting this fact, the author is still conscious that it is difficult to disregard the firm K/Ar evidence as opposed to the less satisfactory fission track evidence, and an average age of 0·61 million years seems virtually established as the real arrival age of the Australasian tektites.

(c) *Age of Deposition of Enclosing or Underlying Rocks of Tektite Recoveries*

The following is a summary of the data available:

Bediasites: occur in close association with outcrops of Oligocene sandstone, from which they are presumed to have been derived. The age of these beds corresponds closely with the K/Ar age.

Georgiaites: these occur in close association with a Miocene formation, and this is believed to have shed the tektites. The disparity between the age of this formation and the K/Ar age suggests that these tektites were redistributed by the processes which formed these later sediments—they are 'reworked'.

Moldavites: these occur in Miocene gravels, the age of which is in close agreement with the K/Ar age.

281

Ivory Coast Tektites: these occur in Quaternary alluvial deposits, for which there appears to be no precise date. This provenance is not in any way incompatible with the K/Ar age dating results.

Australian and South-east Asian Tektites: Philippinites occur in red laterite covered by Middle Pleistocene volcanic tuffs. In view of the uncertainties of Pleistocene time rock divisions, this is in no way conflicting with the K/Ar age. Billitonites occur in Quaternary gravels and tuffs; indochinites in Pleistocene gravels and tuffs, and Lacroix considered that they were about the same age as these host formations—a conclusion which is strongly supported by the K/Ar ages. It is important to note that the Pleistocene (or Glacial Period) is now widely accepted as being much longer than was originally thought, and, while the abrupt climatic change heralding it commenced about three million years ago, the five glaciations (Donau, Gunz, Mindel, Riss, and Wurm) did not commence until after the Villafranchian, not much more than a million years ago. As the Villafranchian is lower Pleistocene, it is clear that what has been called Lower Pleistocene accounts for the greater part of the span of the Pleistocene according to modern thinking, and the Middle Pleistocene extends up to half a million years ago or less. Thus, the tektites of the south-east Asian occurrences probably arrived in what is known as the Middle Pleistocene, but well up towards the top of the Pleistocene epoch, according to modern thinking. This of course applies to the australites also, but it is especially important to realise that terms such as Middle Pleistocene, applied many years ago to south-east Asian host formations, do not nowadays mean the middle of the Pleistocene time span as we now envisage it.

Australites are mainly found in recent soils, recent salt lake fills, dune sands, and fluvial gravels. There does not seem to be any certain case of a deposit of other than recent age yielding australites. Gill has suggested that the Port Campbell, Victoria, australites are in situ, within very recent deposits, a conclusion in agreement with Baker's conclusion that these australites fell only about 5,000 years ago. This evidence of Dr Gill is very carefully documented, and difficult to discount. Yet some of the Port Campbell australites give fission-track ages conforming to the average K/Ar age for australites, and all the K/Ar ages on the Port Campbell australites similarly conform. The Port Campbell australites as a geographical group are *all* astonishingly fresh and undegraded by terrestrial processes, as far as is known, and there seems little possibility that more than one shower

282

is represented at Port Campbell. Hence, it is very difficult to accept Gill and Baker's conclusions, unless the K/Ar age date means something else than is commonly believed.

Petrography of Tektites

The petrography of tektites is relatively simple, the two most important features being the absence of crystallites—imperfectly formed, submicroscopic crystals of equilibrium mineral phases, commonly displayed by volcanic glasses, though some volcanic glasses do not even display crystallites: and the absence of xenoliths, residual fragments of partly melted and assimilated rock, such as are to be expected in impactites. The refractive index of the glass varies slightly from group to group, and within groups (being one of the properties used for very detailed studies of population variations within the individual strewn fields, such as have been carried out by Dr Virgil Barnes and Dr Dean Chapman). This variation reflects silica content, a curve very close to a straight-line plot being produced by plotting silica percentage against refractive index. Moldavites, the most silicic tektites, have the lowest refractive indices, and Ivory Coast tektites, the least silicic, the highest. The cryptic glasses from the Libyan Desert and Mt Darwin, Tasmania, have refractive indices of 1·475 and 1·46, much lower values than any tektites. Specific gravities plotted against refractive indices yield a straight-line plot, with systematic variations between strewn fields. Libyan Desert glass (2·21) and Mt Darwin glass (2·28–2·29) again have lower values than the tektites, which range from 2·30 for moldavites to 2·51 for billitonites.

The Chemistry of Tektites

Chemical studies provide some of the most important evidence bearing on their origin: that is to say, they introduce many additional limiting factors. Major element analyses are given in Table 24 (after the review of this evidence published by C. C. Schnetzler and W. H. Pinson).

The results may be broadly summarised to show the principal variations between tektite groups:

1 *Bediasites*	MgO low
	CaO low
Georgiaites	similar
Martha's Vineyard	similar

2 *Moldavites* SiO_2 high Al_2O_3
                      Fe_2O_3 + FeO $\Big\}$ low
                            Na_2O

3 *Ivory Coast*   More like Australasian than older tektites.

4 *Australasian*   MgO high   (These values are to
                  CaO high   some extent sympathe-
                  FeO high   tically variable with the
                                      silica content.)

TABLE 24

Analysis of Tektites

	Moldavites (14*)	Billitonites and Javanites (13*)	Phillip-pinites (21*)	Indo-chinites (36*)	Australites (32*)	North America (6*)	Ivory Coast (3*)
SiO_2	79·01	71·29	70·87	73·09	73·06	78·59	71·05
Al_2O_3	11·09	12·08	13·48	12·60	12·23	12·57	14·60
Fe_2O_3	0·30	0·78	0·79	0·34	0·60	0·44	0·18
FeO	2·15	5·08	4·30	4·78	4·14	3·01	5·51
MgO	1·49	3·16	2·67	2·16	2·04	0·63†	3·29
CaO	2·08	2·95	3·14	2·31	3·38	0·57†	1·67
Na_2O	0·52	1·57	1·41	1·45	1·27	1·34	1·71
K_2O	3·04	2·20	2·31	2·40	2·20	2·13	1·53
TiO_2	0·69	0·79	0·83	0·87	0·68	0·64	0·73
MnO	0·08	0·14	0·09	0·12	0·12	0·03	0·08
Total	100·45	100·04	99·89	100·12	99·72	99·95	100·35
Total iron content	1·87	4·42	3·89	3·95	3·62	2·65	4·41

* Number of analyses averaged.
† Two analyses from Barnes (1940) excluded for these oxides.

The silica percentages of tektites range from 68 to 82, and the nature of the major element chemistry revealed by these analyses is not unlike that of acid igneous rocks, but with a significant deficiency of alkalis. The major element chemistry is said to be consistent with derivation by differentiation of a parental magma or by mixing of more than one igneous rock magma type. The distinct variation between the groups is most important, as it is a limiting factor on theories of origin. A satisfactory mechanism must be one that could result in such systematic differences between arrival groups. The minor variations within Australasian strewn fields have been studied in detail by Dr Chapman and reveal very complex superimpositions of small sub-groups sharing common properties within the strewn fields, and many systematic variations across the strewn fields. Such variation patterns are reflected in the chemistry as well as the physical and morphological properties. Barnes has made similar studies of the indochinites, and recognises a concentric zonation of property variations outwards from the site of the irregular Muong Nong finds; Barnes relates these variations to the terrestrial impact theory which he favours, but Chapman's very complicated patterns for the australites and south-east Asian tektites do not suggest such a regular zonation, although there is a matching of properties of West Australian tektites and philippinites, and East Australian tektites and Indonesian-Malaysian and Indochinite tektites of the western part of the Asian strewn field. Chapman relates these results to complex effects of lunar spallation of individual tektites, his favoured theory.

Tektite Provenance

Selective volatilisation of the alkalis is generally accepted as responsible for the alkali depletion, and, in support of this, it has been recorded that flanges are depleted more than cores, suggesting loss during ablation.

Various parental materials have been suggested for tektites, considering the bulk composition of major elements, and accepting that volatilisation of alkalis is in all probability a secondary, superimposed characteristic. Chao favours igneous rock, Dr S. R. Taylor a mixture of three parts average shale and one part quartz (also remarking that loess is a chemically feasible parent material, though, as Chao points out, loess, being related to the Quaternary glaciation, is geochronologically impossible), Dr J. F. Lovering favours a granophyre parent (an acid igneous rock of restricted occurrence), while Preuss and Schwarcz have expressed a preference for soil. The latter parental material has been discounted by Dr Chao, who has argued that certain characteristics of tektites preclude derivation as impactites from soil layers below the top 75 cm, an impossible restriction in the light of the magnitude of tektite occurrences. He considered that this restriction is only compatible with the discarded lightning theory of origin! Dr Chao, who is by the very nature of his other researches— on the Cañon Diablo crater, Arizona and the Rieskessel, Germany, for example—likely to tend to lean towards any lunar or terrestrial impact theory, finds a great contrast between the chemistry and physical characteristics of impactites, associated with the Arizona crater and the smaller Henbury, Wabar, and El Aouelloul craters. Such glasses have more diverse composition and contain partly melted fragments, besides containing many more vesicles. It is obvious that meteorite incidence, being fortuitous, will occur on all manner of terrestrial surface materials, both soils and bed-rock formations, and the very uniformity of tektite chemistry indicates that tektites cannot conceivably stem from a simple melting of surface materials. It may be that their composition is similar to a mixture of sedimentary materials as Taylor suggests, but the mechanism by which one could invariably produce such a mixture through the agency of a series of megaimpacts on diverse terrains has not been suggested, nor is it likely to be! Chao considers the volcanic rocks rhyolite and rhyolite-obsidian to be the most likely parent material and favours lunar impact spallation from a monotonous surface of this nature and not terrestrial impact spallation. He regards the very low ferric iron content in tektites as significant—possibly reflecting a low oxygen pressure in the atmosphere of the particular body on the surface of which the impact explosions occurred.

The minor element or trace element chemistry of tektites has been studied in detail by many authorities, and the elemental distribution is of a distinctly terrestrial character, another limiting factor to be taken into consideration in theorising. While Dr Baker is inclined to discount the limitations imposed by this evidence, suggesting that all material in the solar system would be likely to have a similar character, Dr Chao regards it as evidence restricting tektite provenance either to the Earth itself or to some other planetary, asteroidal, or satellite body in the solar system, similarly differentiated. He does not discount lunar provenance, but the implication is that the trace element distribution in lunar material is characteristically terrestrial in character. Apollo evidence is to the contrary.

Gas Bubbles in Tektites

Among the most interesting highly refined studies on tektites have been those on the actual content of gas within bubbles in tektites: the moldavites have provided the most well-known evidence. Certain gas bubbles in moldavites, clearly formed in the primary solidification stage, for they occur in the core of the tektite, have been found to contain vapour with a very low pressure equivalent to the terrestrial atmosphere 32 km up. Argon from these bubbles shows $^{40}Ar/^{36}Ar$ ratios that are typically atmospheric. Gas trapped in this environment must surely either be derived from a previous passage through the upper regions of the terrestrial atmosphere, during which passage primary solidification occurred or else it must reflect the atmosphere of the extra-terrestrial body from which the tektite was derived: a body with a very low oxygen pressure in its atmosphere. The argon isotope ratios have been taken by many authorities virtually to establish terrestrial provenance for tektites.

Cosmic-ray Induced Isotopes

If tektites had orbited in Space for a long time before arrival on Earth they would have been bombarded by cosmic rays for sufficiently long to produce significant amounts of certain isotopes, not otherwise likely to be present because of their relatively short half-lives. Of these radiogenic isotopes, ^{26}Al is the most suitable for measurement because its half-life is neither too short (some of these isotopes have half-lives measured in hours!), nor is it too long. Tektites show a marked deficiency of this isotope, which is considered by many authorities to restrict their provenance to no further out in Space than the Moon. However, a minority school of thought suggests that the actual values do indicate an extra-terrestrial source and one further out in Space than the Moon, so this evidence is slightly ambiguous.

Aerodynamic Considerations

Dr Chapman's experiments do at least show the manner in which flanged australites of button type were formed, and these experiments certainly marked one of the major advances in the understanding of tektites. Yet it must not be overlooked that an entirely satisfactory theory for tektite genesis must explain the little-ablated tektites from other than Australian strewn fields and the quite irregular lumps of tektite glass from Muong Nong, Laos. It is quite valid, however, to take the flanged button as the ideal tektite form, so as to find out by means of simulation experiments the limitations imposed on atmospheric ablation by the nature of these forms. Much mathematical work involving computer programs has, in fact, followed Chapman's ablation experiments, in an attempt to establish how such objects could enter the Earth's atmosphere and ablate in just this fashion. Unfortunately, there is a lack of agreement among scientists versed in ablation mechanics concerning the nature of the limitations imposed. On the one hand, Chapman is quite satisfied that his computations favour lunar-impact spallation of individual tektites, and he even goes so far as to suggest that they indicate a source for the australites in the most prominent of all the lunar rayed craters, Tycho! On the other hand, Dr Adams and his

287

associates disagree with these conclusions: they find that there is only a very narrow range of possible atmospheric incidence, covering a bare two degrees of arc, close to the skipping limit—that is close to a tangential approach to the boundary of the Earth's atmosphere. They also find that some forms of australites are only reconcilable with an entry velocity too low for lunar provenance. In addition, they make the point that the greater part of lunar impact-explosion spallation products would not fall directly on the Earth but would go into Earth orbits, falling to Earth intermittently, just like true meteorites, over a long period, as these orbits decay, and having a distribution over a much larger arc of the globe than the tektite strewn fields —the entire globe is likely to be sprinkled with these delayed arrivals.

Adams and his associates favour a parent body spallation theory. The theory of lunar parent body spallation involves a large parent body grazing the edge of the Earth's atmosphere at about escape velocity (11·2 km/s) and shedding numerous ablation drops, some of which exited from the atmosphere again with a velocity a little less than the escape velocity. Solidification took place in Earth orbit, and this orbit led them to entry into the Earth's atmosphere over Australia. Adams believes that this theory counters his own objections to Chapman's theory, but it seems to the writer that it fails to counter one objection, based on the lack of near-miss spallation products behaving like true meteorites. The theory seems to require an extraordinarily special set of circumstances.

Adams and Chapman both agree that material could not be ejected from the Earth either as a parent body or a molten spray from which individual tektites solidified on the way up. Chapman produces arguments showing that rigid spheres could not overcome the effects of aerodynamic pressure, except in quite untenable circumstances. These conclusions are absolutely critical in respect to either the theory of terrestrial impact spallation (favoured by Barnes) or the theory of terrestrial cryptovolcanic projection (which the author finds not entirely unattractive); though, as will be apparent in Chapter 22, there is something to be said for both theories, the aerodynamic evidence against terrestrial provenance in any form, advanced by Chapman and Adams, would have to be broken down before these theories could be accepted. It has been suggested by some authorities that this aerodynamic evidence could be broken down, but it is difficult for any but specialists in this field of research to make such critical evaluations. However, all those who have attempted to solve the riddle of the tektites have been prepared to reject some evidence in someone else's specialised field, and no one is exempt from this criticism!

It must be concluded that, on the face of it, the ablation evidence is against any form of terrestrial provenance, somewhat against particulate provenance from the Moon, and reconcilable with parent body provenance from the Moon, but only by invoking very specialised limiting conditions, while for other reasons such a theory appears untenable.

Relationship to Other Occurrences of Cryptic Glasses and Craters

From Table 23 it can be seen that there are four remarkable coincidences between:

(a) The potassium/argon and fission track ages of the geographically adjacent australites and Mt Darwin glass, Tasmania.

(b) The potassium/argon and fission track ages of the moldavites and the cryptic glass (suevite) from the geographically adjacent Rieskessel crypto-explosion structure, South Germany.

(c) The potassium/argon ages of the Ivory Coast tektites and the cryptic glass from the geographically adjacent Bosumtwi crater, Ghana.

(d) The potassium/argon ages of bediasites and georgiaites and the rather distant Clearwater Lakes craters in Quebec, Canada, and also the very distant Libyan Desert glass occurrences.

These coincidences cannot, surely, be entirely fortuitous, but what exactly do they mean? How much weight should be given to them in our search for a common origin of tektites? The chemistry, physical properties, and petrography of Mt Darwin glass does nothing to support congenetic relationship to australites. The nature of the Australasian strewn field is such that a single source in Tasmania is quite inadequate. There are great difficulties in genetically relating the moldavites and the Rieskessel on an impactite hypothesis: in addition to the potent arguments advanced by Professor W. H. Bucher in favour of a joint endogenous, cryptovolcanic origin for the Rieskessel and the nearby Steinheim Basin (Chapter 20), Professor von Englehart has concluded that the only possible mode of derivation of moldavites from the Ries crypto-explosion is by fusion and homogenisation of a deep block of country rock beneath the focus of the structure (the site of impact). He states that there is no possibility of glasses of the chemical composition of the moldavites stemming from simple fusion of the visible country rock. In addition suggestions by Cohen of a progressive variation of moldavite properties outwards from the Rieskessel have been refuted—indeed all the evidence now suggests that the source of the moldavites would have to be to the east of the strewn field, in quite the wrong direction. The study of Schnetzler, Philpotts, and Thomas of the Ivory Coast tektites–Bosumtwi crater relationship is equally suggestive of some genetic relationship, but not perhaps the direct one popularly favoured. Working on rare-earth and barium trace element abundances they find some resemblance in their geochemistry between Ivory Coast tektites and the country rocks of the crater (metasediments), but no exact match. On the other hand, the cryptic glasses associated with the crater give a remarkably good match. The potassium/argon ages of tektites and glass are similar, according to Gentner and associates, and Zahringer; and similarities are also found in other isotopic relationships. The conclusions of Schnetzler and associates cannot be given in detail here, but their work does seem to establish that there must be some connection between tektites and cryptic craters or crypto-explosion centres. These authorities suggest the derivation of the tektite glass by deep-seated homogenisation and fusion of a large block of the crust beneath the Bosumtwi centre, independently suggesting a process very like that suggested by von Englehart as the only possible mode of impact explosion derivation of the moldavites from the Rieskessel.

Chao, however, as has been noted, has come out strongly against terrestrial impact explosion as a source of tektites; and so, basing their work on ablation studies, have Chapman and Adams. Chao's preference for an igneous parent might be equated with a deep-seated fusion and homogenisation hypothesis, but the doubt still remains whether even such a process would result in such chemical monotony as is possessed by the tektites. The lunar theories of Adams and Chapman are quite incompatible with any relationship between tektites and large-scale crypto-explosion structures on the Earth's surface— if this relationship is not fortuitous, Adams' and Chapman's theories are untenable: it is as simple as that. A relationship between tektites and crypto-explosion structures is not completely incompatible with lunar tektite provenance, but the theory that could reconcile the evidence is so complex that it need not concern us further.

The derivation of the Bosumtwi crater glass from the country rocks is a remote possibility, but the derivation of the Rieskessel suevite from the country rocks is inconceivable: and so, as the only possible alternative theory to deep-seated fusion and homogenisation (the only possible mode of derivation of tektites as terrestrial impactites) we may entertain a crypto-volcanic theory. There is, in fact, strong evidence that the Bosumtwi crater is crypto-volcanic, in an excellent paper published some thirty years ago by N. R. Junner. It is very difficult to reconcile the youthful age of this thirteen-kilometre diameter structure with impact explosion, for there is a complete dearth of products of explosive evisceration outside the crater: where could it have gone? It seems likely that, notwithstanding the fact that breccias associated with cryptic glass, and clearly related to explosive effects, are associated with the structure as irregularly distributed outlying occurrences outside the actual crater, the crater itself may well have been formed by a post-eruptive subsidence mechanism characteristic of volcanic caldera formation. Similarly, there is reason to believe that the Rieskessel structure had a two-stage development (irreconcilable with the single-stage impact explosion process), the stages being appreciably separated in time so that they are stratigraphically distinguishable: the existence of ultrabasic igneous rock beneath the structure is suggested by the geophysical evidence, and there is a geotectonic lineament along which are situated the Rieskessel structure, the Steinheim Basin (another such structure), and 'volkanembryonnen', a nearby cluster of volcanic diatremes. The Rieskessel, like Bosumtwi crater, is by no means an established impact explosion structure, to say the least!

The last pair of coincidences in potassium/argon dating also seems to favour an endogenous origin for the cryptic structures and tektites, if these coincidences mean anything at all. The Clearwater Lakes double crater structure is by no means a well-established meteoritic structure. Currie and Shafiqullah have brought forward geochemical evidence relating this structure, and other Canadian structures of similar character, to alkaline volcanism, of a type commonly associated with diatremes—gas drilling vents of explosive character. The cryptic glass of Clearwater Lakes seems to be best interpreted as a product of melting of country rocks by extremely hot, gaseous eruptive agents, beneath the structure. Some of the supposed impactites have the character of igneous rocks and there are even ultra-basic rocks represented! These glasses have little other than the potassium/

291

argon dating to connect them with tektites, and the bediasite strewn field is extremely distant, many hundreds of miles away. To derive the bediasites and georgiaites from this structure verges on a fanciful suggestion, and so, perhaps, the age coincidence only means that a number of related events took place in this general area of North America at that particular time. Crypto-volcanic events might well have been spread over a major province of large extent, in which many of the structures were contemporaneous. If the distance between the Clearwater Lakes structures and the bediasite strewn field is immense, that between the Libyan Desert glass occurrence and the bediasite strewn field is so immense that the coincidence of age-dating values can surely only indicate some sort of broad connection as is suggested above. The relationship of tektites to cryptic glasses and crypto-explosion structures associated with them may well, then, be a real one, but scientists should beware of drawing the conclusion that this relationship is an indication that the tektites are terrestrial impactites. There is at least more reason to bring forward this evidence in favour of a terrestrial crypto-volcanic origin for tektites.

It might be profitable, taking into account these coincidences of ages and the distinct geochemical differences between the four age groups of tektites, to accumulate evidence on geographical and geochronological distribution of geochemical major element variation in acid intrusive igneous and volcanic rock suites, to see if any correlations are evident. As far as the author knows, little geochemical work relating to tektite genesis has ever been done on these lines.

Finally, in connection with the age-dating coincidences, it must be noted that there is some suggestion that the crypto-explosion structures giving coincident ages with the tektites only form part of larger geographical 'provinces' of such structures, including individual structures of ages ranging from Palaeozoic to quite recent. In contrast to these 'provinces', large areas of the land surface of the Earth appear to show no such structures. Such a distribution pattern for crypto-explosion structures, if real, would be very like the pattern of carbonatite diatremes, which, in Africa, form a geographical province broadly related to the Great Rift Valley anorogenic zone, but are of ages ranging from Precambrian to Recent. There certainly does seem to be a Mississippi-Eastern Canada province, a north-west African province, a north-east European province, and an Australian province of such structures, of varied ages, and there also seems to be some correlation between this distribution pattern and the distribution of tektites. But this may all be an illusion—it is certainly very easy to imagine such distribution relationships!

22: The Theories of Origin of Tektites

'*Where they come from no-one knows*' (A. S. Woodward, 1894)

Three-quarters of a century after the above dictum was pronounced it remains as valid a conclusion as ever. All one can do, to round off this review of tektites, is to enumerate the several theories, and then try to give some weight to the evidence favouring and countering the several not quite untenable theories.

The names reportedly used by the Indomalaysian peoples for tektites—'sunstones', 'stardung', and 'moonballs', etc—certainly call to mind extra-terrestrial provenance, but such names probably have no bearing on their provenance, for it is almost certain that no one has ever seen a tektite arrive on Earth.

The earliest theories invoked artificial origin; from Darwin's time to the end of the nineteenth century volcanic origin was favoured; and then during the last seventy years, extra-terrestrial origin has been favoured—though terrestrial origin has never been disproved. The theories so far advanced are listed below:

1 Artificial origin

Furnace slags	
Gas works slags	Lindaker, 1792
Burning of earth	
Accidental artifacts	Hillebrand, 1905
	Berwerth, 1917
Purposeful artifacts by 'savage'	
or civilised man	Hillebrand, 1905
Tin slags	de Groot, 1880

The artificial origin theories have to some extent found favour because artificial glass fragments have been confused with tektites. The discovery of the age of some tektite groups (eg, bediasites, georgiaites, moldavites) has entirely discounted theories of artificial origin.

2 Natural fires

> Burning plant material
> Ignition of coal seams

Neither the chemistry nor the physical properties of tektites are compatible with such an origin, and the ablation characteristics of australites entirely discount this theory.

3 Concretionary in limestone   Jensen, 1915

This is of purely academic interest: like the theory above it is quite untenable on account of ablation characteristics.

4 Abrasion

> Water worn, rolled, abraded and
> shaped obsidian pebbles
> Wind blown sand-eroded obsidian
> fragments   Merrill, 1911
> Shaping in the gizzards of emus

Likewise, these theories are now of purely academic interest.

5 Dessication

> Drying up of silicate gel masses   Wing Easton, 1921
>   Van Ider, 1933

This is an entirely fantastic theory, without an element of truth according to Dr George Baker, and no one nowadays would surely disagree with him!

6 Lightning

> Fusion of dust particles in the Earth's
> atmosphere   Gregory, 1912
>   Chapman, 1929, 1933
> Fusion of sand on the ground

These theories do have some reason for support, in the known occurrence of glass tubes known as *fulgurites*, which are believed to be produced by lightning strikes. The 'aerial fulgurite' theory of Gregory and Chapman is strikingly similar to early theories concerning true meteorites (Chapter 2). The nature and scale of tektite occurrences as now known completely discounts the lightning theory, in any form.

Fulgurite glasses are, incidentally, quite unlike tektite glasses.

7 Terrestrial volcanic origin

Moldavites 'glassy lavas'	Mayer, 1788
Billitonites 'obsidian'	Wickman, 1893
	van Dijk, 1879
Australites 'obsidian bombs'	Charles Darwin, 1844 and many other nineteenth-century writers
Krakatoan provenance for australites	Simpson, 1902
New Zealand provenance for australites	
Australites 'volcanic ejectamenta'	W. D. Campbell, 1906
'Lava spherulites'	Lawson, 1902
'Bursting lava bubbles'	La Conte, 1902
'Relationship to Pele's tears'	Moore, 1916
Bubble hypothesis	Dunn, 1908, 1912
Rain, hail, or snow falling on molten lava	Dunn, 1935
Bubble hypothesis	Buddhue, 1940

Tektites have compositions which are unlike those of any known terrestrial volcanic rocks. As has been noted, in the last few years it has become apparent that there is some similarity between tektite glasses and glasses associated with cryptic structures of explosive origin, or closely associated with explosive brecciation phenomena (eg, Bosumtwi crater, the Rieskessel, Clearwater Lakes). However, whatever the origin of these glasses, they do not appear to be pure igneous melt solidifications (evidence of Currie and Shafiqullah, 1968), and can only be either crypto-volcanic products, combining volcanic emanations of an unusual nature and products of country rock fusion by the agency of these hot emanations. While it is highly likely that there is a crypto-volcanic extension to the generally accepted range of volcanic processes, in all probability most closely related to alkaline diatreme (carbonatite, lamproite, kimberlite) volcanism, the author believes that these processes should be regarded as crypto-volcanic rather than volcanic, and he does not see much support for a pure volcanic theory for tektites.

Terrestrial Crypto-volcanic Origin

This theory has, apparently, never been concisely formulated, but, with the arguments advanced in the last few years by Bucher, Currie and his associates, McCall, Snyder, and Gerdemann, Amstutz, and others, it seems that there is room for such a theory, involving direct genetic association with structures like those of the Rieskessel, Bosumtwi crater, and Clearwater Lakes craters, the glass being the product of a combination of primary volcanic emanations of a volatile nature and melt products from the country rocks as suggested above (following Currie and Shafiqullah's arguments). There is no real difficulty in deriving tektite glasses this way, but the mode of development of the primary forms, the incorporation of low-pressure atmospheric gases in bubbles, the ablation characteristics and above all the strewn field distribution patterns would all have to be fitted into such a theory. The fact that the strewn fields appear to be composite, as has been shown by Barnes for the indochinite field, and by Chapman for the Australasian field as a whole, could be reconciled with such a theory. Its virtue is the fact that it obviates the difficulty involved in genetic relationships between the crypto-explosion structures and the tektites, and fits in with evidence now coming forward supporting Bucher and the present author's arguments that these structures are of endogenous origin. Its greatest weakness is the difficulty in getting the tektites out through the Earth's atmosphere (stressed by Chapman and Adams). But, in spite of this, it remains the best of the terrestrial theories.

Primaeval Terrestrial Heating and Expulsion Theory

This theory due to Kuiper (1953, 1954) invokes silicate material expelled from the Earth by the evaporation of an appreciable fraction from its molten surface during the peak of post-accretional heating up of the entire planet. This material was supposedly ejected into interplanetary space and acquired a circumsolar orbit, was again melted and degassed during a close approach to the Sun, and finally ultimately collided with the Earth, producing the secondary ablation modifications of the tektites. This theory is similar to one of the theories involving acid meteorite parent material from the asteroid belt, only differing in that the material of tektites has an ultimate particulate, terrestrial, rather than an asteroidal, provenance.

Urey's evidence against circumsolar orbits for particulate tektite clusters, and the evidence of entry velocities and cosmic-ray-induced isotope abundances, are all against such a theory, which is of a very hypothetical nature.

Terrestrial Impactite Theory

This theory combines extra-terrestrial agencies with a terrestrial derivation for the actual glass.

It has various forms:

Cosmic body collision	Spencer, 1933
Impactite theory	Urey, 1955, 1957
	Barnes, 1961
	Cohen, 1961
	Gentner, Lippolt, and Schaeffer, 1962

Though at first sight attractive, this theory encounters so many difficulties that it has been abandoned by many of those scientists at first inclined to favour it. The difficulty of getting the material out from the Earth in the first place is shared with the crypto-volcanic theory: atmospheric retardation of hypervelocities is tremendous and consequently restricts the flight range of small objects. Tektites as individuals are almost certainly incapable of escape, and some theory involving a parent body, gas, and droplet blast, or a molten jet, is perhaps the only alternative. Though Chapman and Adams rule out terrestrial provenance for tektites, in any manner, perhaps less is known about such processes, and summary rejection of them might yet prove to have been premature. The possibility of atmospheric blow out in the plume over an impact explosion crater such as the Arizona crater has been suggested by Urey and others, but Chapman finds the aerodynamic objections to such a process overwhelming in the case of individual tektites as rigid glass objects.

The distribution of tektites in the Australasian field is such that no single impact focus could have distributed them all: multiple sources would have to be postulated (this, however, would be in agreement with the complicated population patterns which favour individual sub-provinces within the vast Australasian strewn field).

The monotonous chemistry, the major element chemistry variations between groups and the lack of partly fused xenoliths are all characteristics strongly countering the impactite theory. The only possible form of this would seem to be one invoking deep seated homogenisation of a block of country rock beneath a mega-impact focus

(Chapter 20)—yet there are many objections to this, and the author regards the terrestrial impactite theory as a poor second behind crypto-volcanic theory, while admitting that these two of all the theories invoking terrestrial provenance for the tektite glass are the only two remotely tenable. The 'meteoritic' spheres (of kamacite, troilite, and schreibersite) in some philippinites could be explained by such a theory—as incorporated traces of the impacting meteorite—yet why should such metallic spheroids be so scarce if this is their origin?

There is no need to distinguish cometary from meteoritic terrestrial impact theories, for the objections remain virtually the same. Another form of impactite theory invokes 'antimatter' ('Dirac-style matter'), either cometary or meteoritic: such explanations have lately been advanced to explain the Tungus event (Chapter 20) but, though the novelty of this idea is attractive to many scientists, there is really little or no evidence to support it and some valid evidence against it. Advanced as an explanation for tektites by Khan in 1947, it meets as many objections as the orthodox versions of the impactite theory developed from the original meteoritic splash theory of Spencer.

Meteoritic Origin of Tektites

The meteoritic origin of tektites is really the most simple and obvious of all theories, taking into account their ablation characteristics. Theories deriving them as shed melt during the ablation of meteorites were favoured by Lacroix and Suess in 1932. It is quite impossible to derive them from the shed melt of stony meteorite types known to us, for the composition of the fusion crust is that of the ultrabasic or basic meteorite from which it is derived, and the glass produced is heavily charged with magnetite dust. Lacroix and Suess could only refer the tektites to some unrecognised form of acid meteorite, for the existence of which there is no evidence whatsoever.

Direct derivation from an acid or granitic meteorite of a type beyond the range so far recognised in falls and finds is the basis of theories favoured by Goldschmidt in 1921 ('granitic kosmolite'), by Linck in 1926, and by Paneth in 1940. Paneth suggested melting of an acid meteoroid body at perihelia by the Sun's heat, the glass being converted to discrete droplets which are the individual tektites, later ablated on atmospheric entry. Baker has found neither the chronological nor the geographical distribution of the tektite groups to favour such a theory, but does regard the theory as compatible with

300

the presence of inclusions of lechetelierite within tektites. Recently, such inclusions in indochinites have been regarded as derived from quartz grains, probably not terrestrial xenoliths. The greatest objections to meteoritic theories are the lack of any record of acid meteorites, the almost complete lack of similarity mineralogically or chemically between tektites and meteorites, and the different Earth-arrival patterns (tektites coming not continuously but in four distinct geological events).

The meteoritic theories do, however, find strong support in the presence of kamacite, troilite, and schreibersite in metallic microspherules within some philippinites. These minerals are either rare or completely unknown in terrestrial rocks, while they are the common stuff of the siderophile and chalcophile fractions of meteorites.

The chemistry and mineralogy of some meteorites—for example the eucrites—is very like terrestrial igneous rocks (dolerites, gabbros) and the geochemistry of meteorites is not entirely unlike the terrestrial geochemistry. There is no doubt that the derivation of tektites from a strange form of rare visitant meteoroid remains one of the remotely acceptable theories.

The Cometary Origin of Tektites

This was seriously advanced by H. E. Suess in 1951, and remains by no means the least tenable theory. Urey has supported this theory. The comets, composed of ices, are supposed to have lost their volatiles at perihelion, near the Sun, and the depleted residue produced the tektites. The oxygen isotope chemistry of tektites is not adverse to this theory, but the presence of lechatelierite and the total absence of ammonia, a major component of comets, are against such an origin. Comets are now widely held to be members of the solar system, so this theory does not imply extra-solar system origin, though it is a conceivable corollary of the theory.

The chronological distribution and geographical distribution of tektites seem to be against this theory, for comets would be expected to be regular visitants throughout the geological record and to sprinkle the whole surface of the Earth. The resemblance of the geochemistry of tektites to that of igneous rocks is also adverse evidence.

Another cometary theory is favoured by Lyttleton. This involves derivation of tektites from a jet directed vertically downwards from the tail of a comet passing through the edge of the Earth's atmosphere. Lyttleton regards low eccentricity comets as the likely source, and considers that they could accrete material from their tails during a passage lasting a few hours and project it vertically downwards through the Earth's atmosphere. This theory fits in well with the low latitudes of the tektite strewn fields and their restricted extent, in the case of such comets. Most materials would be vaporised, and, again, a residue of the more refractory materials is supposed to produce the tektite material, in this case within the jet stream. He considers the speed of entry would satisfy the requirements of the ablation characteristics of australites. The dimensions of the strewn field would reflect rotation of the Earth during the short period of cometary passage. The total mass of tektite material in the four groups is said to be of the right order for this to be an acceptable theory.

If either of the cometary theories is correct, there can be no connection between tektites and terrestrial crypto-explosion structures: the potassium/argon dating and geographical coincidences must be entirely fortuitous.

Asteroidal and Planetary Theories: Satellites Other than Lunar

There are a number of such theories, which are not entirely unconnected with the meteoritic theory of extra-terrestrial provenance. Of these may be listed:

Derivation from a satellite of the Earth 'X' which fell to Earth in Tertiary times	Belot, 1933
Derivation from the Sial (siliceous crustal layer of the continents) of a planetoid	Denaeyer, 1944
Original glass skin of a cosmic body	Washington and Adams, 1951
Lost planet between Mars and Jupiter, similar to the Earth but with a glassy surface layer, later disrupted	Stair, 1964
A self-melting meteoritic planet, disrupted quite recently, due to concentration of radioactive minerals in the acid crust	Cassidy, 1956
Celestial body destroyed by collision, containing sediments and igneous rocks just like the Earth	Barnes, 1957

These theories are very similar: they really represent attempts to find a source for a postulated acid meteoroid material from which tektites have been supposedly separated by one of a variety of processes, and later modified by atmospheric ablation. Such theories may involve direct separation of tektites or separation at a later stage— the latter being an acid meteoroid parent body hypothesis. Such parent body hypotheses may invoke melting off at perihelion or on incidence with the Earth's atmosphere, but prior to ablation. Chapman finds some such theories to fit his requirements for a negligible aerodynamic pressure differential (between stagnation point and base) during primary formation, but regards the undetectable ^{26}Al content as imposing a restriction of the cosmic flight time to less than 90,000 years (quoting Viste and Anders), a value too low for provenance in the asteroidal belt. Meteoritic incidence velocities to the Earth's atmosphere are too high according to Whipple and Hughes—about 17 km/s, significantly greater than deduced australite entry velocities.

There are arguments advanced by Professor Urey concerning requisite cluster densities needed to prevent tektites on a heliocentric orbit, such as meteorites from the asteroidal belt would possess, being dispersed by perturbations due to the Sun's gravitational field: such arguments are taken to require that tektites of asteroidal or cometary provenance would be piled a metre deep in the case of the Australian strewn field. While Baker brings forwards arguments based on the terrestrial dissipation of tektite material to counter this argument of Urey's—invoking abrasion, weathering, etc—it seems unconvincing, and Urey seems to have raised a very valid objection to derivation of tektite swarms as orbiting clusters from the asteroidal belt or from comets. His objection would apply to any theory invoking melting at perihelion, but not to a theory invoking shedding of individual tektites from a parent body at the time of atmospheric incidence and prior to ablation.

Lunar Impactite Theories

There are two modern versions of the lunar spallation theory, one advanced by Chapman involving spallation of tektites as individuals, and the other advanced by Adams and Huffaker, involving the spallation of parent bodies. This hypothesis was first advanced by Nininger in 1940: he pictured bombardment of the Moon by meteorites causing violent splashes of lunar rock to be projected upwards, and out of the gravitational control of the Moon. The absence of a lunar atmosphere would allow impact of meteorites at unabated cosmic velocities because there is no atmosphere to brake them in their approach, and it would also facilitate escape, even of small objects such as tektites.

This theory is popular in America, partly, perhaps, because of the somewhat emotional emphasis placed there on the meteoritic theory of lunar cratering, even though many authorities regard it as unsound—Spurr, Green and Poldervaart, Moore, Fielder, and McCall to name a few (see Chapter 23). Adams and Huffaker regard the computations of Chapman, which support a lunar origin for the australite flanged buttons, as incorrectly derived, and using a new value for one of the quantities involved in these ablation studies, show that the conditions are much more restricted than Chapman supposes. The entry velocities of some tektites are too low for the lunar provenance they suggest, and the angle of entry confined to a 2° arc near the skipping limit (Chapter 21). Adams and Huffaker thus invoke a parent body, of lunar spallation origin. Their theory is somewhat similar to other parent body ablation theories producing tektites as melt drops favoured by Hardcastle (1926), Lacroix (1932), Fenner (1938), and O'Keefe (1960). Kuiper (1954) has also favoured lunar origin, which he regards as compatible with the low age of tektites (K/Ar), the low bubble pressures, the grouping in showers, and peculiar, monotonous composition range, unlike that of meteorites and characteristic of products of a high-temperature fractionation process.

There are certain objections to lunar origin that are shared by all versions of lunar impactite theories.

(a) There is much evidence to suggest that the lunar craters are not of explosive origin, and are not astroblemes but products of some endogenous process.

(b) If, say, Clavius (228 km diameter) and the maria (several hundred kilometres in diameter) *are* astroblemes, then surely they would have hurled enormous masses of rock into Space, not just objects measured in grammes! Chapman's 'tychtite' theory is surely the classic case of the mountain 'Tycho' giving birth to a mouse!

(c) The accruing body of analytical data from the lunar surface material show it to be basic, not acidic. It has a peculiar geochemistry, quite diagnostic.

(d) Where are the near-misses that would have gone into decaying Earth orbits and fallen to Earth intermittently over a long period just like meteorites? Not one trace of lunar material with its high titanium and peculiar trace element characterisation has even been found on Earth!

(e) If the Moon produced the tektites from Tycho, or any other crater, for that matter, *as little as two-thirds of a million years ago*, could it have settled to its present quiet (almost moribund?) state, in such a short interval of geological time as has elapsed since? It is extremely unlikely, considering the Apollo age-dating results, that Tycho is as young as this.

(f) There are only four age groups of tektites: the number of lunar craters is measured in hundreds of thousands! Tektites should be a prominent feature of the rocks throughout the geological record if their production by the agency of lunar impacts continued right up to Quaternary times.

(g) The probability is that *all* the large scale lunar cratering, whatever its origin, took place long before the arrival of the first known tektites on Earth.

(h) The rubidium-strontium dating is incompatible with lunar origin (though, admittedly, it is incompatible with any theory so far propounded!).

(i) The theory seems to require a thoroughly differentiated Moon: considerations of the nature of the lunar surface and that of the Martian surface revealed by Mariner IV suggest rather that these bodies retain primitive surfaces reflecting a lack of *internal differentiation*, a conclusion which there is much other evidence to support on other grounds. What limited lunar differentiation there was, though widespread is of a kind not likely to produce tektite material as an end product.

(j) The geochemistry of lunar samples is not like that of tektites and no such acid rocks have been recognised on the lunar surface.

Lunar Volcanic Theories

First suggested by Verbeck in 1897 and revived by Linck in 1928, this theory encounters most of the objections raised to the lunar impactite theory. While there is some swing at the present time towards a volcanic interpretation for much of the large-scale lunar cratering, there is also much evidence to suggest that subsidence of a volcano-tectonic nature and not explosive evisceration was the mechanism by which the craters formed. The nature of the lunar surface does not seem to fit this theory too well, but it is just conceivable that acid ejecta could have been explosively produced, even though the surface is of mainly basic composition. The objections raised on the basis of chronology, lack of near misses, the present quiet or moribund nature of the lunar surface as well as the anomalous rubidium-87/strontium-87 dating and strontium-87/strontium-86 ratios seem to relegate this theory, like the impact spallation theory, to an almost academic status.

Solar Prominences

A hypothesis involving periodic variability of the Sun's activity, related to attempts to explain the Ice Ages, was advanced by Himpel in 1938. Nebulous material, projected from solar prominences and approaching the Earth, has been supposed to produce clouds from which hydrogen and a small residue which formed the tektites are supposed to have separated. As there is now known to be no synchroneity between tektite showers and Ice Ages, the whole 'raison d'être' of this theory has been lost!

Extra-solar-system Origin

Orginally suggested by Krause in 1898, this theory was revived by Kohman as late as 1958. The entry velocities required on astronomical arguments are far too high to be applied to australites, taking into account their ablation characteristics. The cosmic-ray-induced isotope content (^{26}Al) does not support this origin as such an origin would require the tektite material to be subjected to interstellar cosmic radiation over a very long period according to Viste and Anders.

Conclusion

After hearing the author lecture on tektites, a friend of his—a geologist and philosopher, Dr George Seddon—remarked, 'before hearing you lecture I thought tektites were quite incredible, now I know they are impossible'. Woodward voiced a similar scepticism concerning explanations available in his day: with the increase in the volume of research, the introduction of sophisticated and highly refined instrumental research methods, and studies of innumerable side-issues to the real problem, the scientist of the nineteen-sixties has increased the number of possible answers and increased the number of anomalies! The conflict of evidence remains baffling and quite unresolved. Concerning tektites, Paneth remarked, echoing Fourcroy: 'by eliminating

306

the absurd or impossible, one finds oneself compelled to adopt what would previously have appeared almost incredible'. Following this precept, the author can only conclude that of all the theories that are remotely possible, but still incredible (!), he finds that:

(a) Lunar impact spallation, lunar volcanism, terrestrial impact spallation, and terrestrial vulcanism are the least credible, if such a statement can be made concerning tektites.

(b) The presence of kamacite, troilite, and schreibersite in some philippinites suggests some genetic connection with meteorites. Some form of direct meteoritic theory not involving clusters of actual tektites orbiting in circumsolar orbits, and not involving derivation from spallation resultant on mega-impact explosions of Earth and Moon or any other solar system surface, seems favoured by this evidence.

(c) A cometary theory is better than (a) but weaker than (b).

(d) Neither (b) nor (c) is reconcilable with a relationship between crypto-explosion structures and associated glasses, and tektites. The author believes that this relationship is virtually established beyond the bounds of coincidence in the case of the Rieskessel and Bosumtwi structures. He therefore leans towards the almost incredible, and suggests that the resolution of the tektite enigma is most likely to lie with the crypto-volcanic theory, with eventual breakdown of the most damaging evidence against it, ie, aerodynamic evidence against projection from and return to the Earth advanced by Chapman and Adams, with their associates.

(e) However, despite the above conclusions, it is noted that the solidification age of tektites is very near the exposure age of meteorite irons (Chapter 18). This suggests that we may, perhaps, yet have to seek a source for tektites related to the break up of the asteroids, the almost certain source of true meteorites, unlikely as it seems. Perhaps, after all, they are unusual forms of true meteorites!

23: The Moon, Mars, and Meteorites

The cratered surface of the Moon first came to the notice of Galileo using the newly invented telescope, and late in the seventeenth century, Robert Hooke made some fine drawings of crater structure, likening the lunar surface to the close cropped uplands of Salisbury Plain in England! He interpreted the craters as volcanic, though mentioning the possibility of the impact of some foreign bodies. Hooke, in fact, lived before the reality of meteorites was scientifically accepted. Thirty years after the publication of Chladni, von Gruithuisen, in 1826, suggested that the craters were the result of meteorite impacts. It was, however, Gilbert, a geologist, who really developed the meteorite impact theory of lunar cratering, as late as 1893. Since restated, over and over again, by Baldwin, Urey, Shoemaker, Kuiper, Dietz, and others—mostly American scientists—this theory has not undergone much development since Gilbert's time. Up to the Ranger missions the only significant development was the recognition of the process of impact explosion, including vaporisation of the impacting mass and fusion of the rocks of the host surface. These studies drew heavily on nuclear explosion experiments producing artificial craters. Impact explosion was found to be the expectation of large impacting bodies falling on earth at high residual cosmic velocities. Small bodies would behave in the same way when impacting on the atmosphere-less surface of the Moon. Mile-scale craters could be produced on Earth this way, and much smaller craters on the Moon. After the successful Ranger 7 mission, scientists in America had to embark on a crash revision of their ideas, and the theory of 'secondary impacts' was born, being stated by Shoemaker within three months of the mission. Time will tell whether this hurried revision produced a realistic working theory or simply held back the tide of contrary evidence. The revised theory invoked secondary ejection of rock masses spalled out of the giant rayed craters such as Copernicus and Tycho, and scattered hundreds or even thousands of miles away from these foci, forming the secondary craters and rays as they, in turn, impacted on the surface of the Moon.

Besides the novel ideas of Gilvarry, who has invoked impact in static bodies of water which supposedly filled the mare depressions, and those of Urey, who has invoked ice melting resultant on cometary impacts to form the sinuous rilles as water scouring channels, little need be added here concerning the impact theory (which is more fully discussed by the present author in his review article, 'the Lunar Controversy', published by the British Astronomical Association), except to add that the discovery by Mariner 4 of the mega-cratered, barren wasteland surface of Mars inevitably extended the application of this theory to the small red planet. And Mariner 9 has lately revealed strange polygonal craters difficult to equate with impact explosion on the surface of the extremely distorted Mars satellite, Phobos.

The theory received a rude jolt from the discovery of alnöite, a rock-type of extreme rarity related to the explosive carbonatite 'diatremes', and closely related to kimberlite, in the Brent crater structure, Ontario, by Currie and a co-worker. This had been considered a type example of an ancient impact explosion crater, and was known to have coesite association (the high-pressure silica polymorph being regarded as the hall-mark of impact explosion), and to display shock lamellation effects. The discovery threw doubt on the validity of the meteoritic interpretation of the Mississippi chain of crypto-explosion structures (associated with shatter cones) and Canadian extension of that chain. Discoveries of eruptive rocks associated with the Clearwater Lakes, Manacouagan Lakes and New Quebec crater structures by Currie and associates have engendered further doubts. Snyder and Gerdemann, and Amstutz, have independently adduced evidence against meteoritic involvement in the formation of the Mississippi structures, regarded as crypto-volcanic by Bucher.

Turning to the Moon itself, Gilbert as long ago as 1893 admitted that chains of lunar craters such as those of the Hyginus Rille could only be of volcanic origin. Barrell and Spurr later adduced much evidence for lunar volcanism, and since 1950, von Bulow, Green, Poldervaart, Moore, Fielder, Katterfield, Erlich, Steinberg, Gorshkov, Leonardi, and the present author have brought forward much additional evidence. Rangers 7–9 produced close-up photography including evidence for lunar volcanism—especially in the form of craterlets situated on and aligned along the summits of abrupt ridges.

Luna and Surveyor missions revealed rocks of fragmental and vesicular appearance, and later Surveyor missions revealed a fragmental layer or regolith a metre or so thick, unfortunately misnamed 'lunar soil'. Alpha particle back-scattering was used to analyse the rocks, indirectly, and the results suggested basalt. Orbiter oblique photography, a spectacular success, revealed many structures of unequivocal volcanic origin. Apollo 11, 12, and 14 recovered actual iron-rich basaltic material *sensu lato*, with the non-volatile element titanium enriched to varying degrees, and taking the place of silicon, as it were, so that some of the lunar basalts tend to have the low silica content of terrestrial ultrabasic rocks (less than 45 per cent). These rocks are, texturally and mineralogically, uncannily like terrestrial basalts and dolerites, and also like the 'basaltic achondrites', when superficially examined. Yet the Moon rocks have a geochemistry of their own—minor element distribution plots from all three Apollo recoveries show a pattern that seems to exclude the derivation of any meteoritic material from the Moon, and also suggest that they could have had no unity with terrestrial material on the Earth, except possibly in its very early formative stages—and even this appears unlikely. In particular, the depletion of the element europium is evident in all lunar rocks, and not in meteoritic or terrestrial material.

Most scientists, even in America, are now accepting that there are primary eruptive rocks on the Moon. The mare-filling basalts seem to have been differentiated, as far as they have been differentiated at all, from a very large volume of magma, a fact that is against production contingent on melting consequent to local impact events.

Scientific conferences dealing with lunar mission results deal at length with igneous petrology, the subject-matter is so like primary terrestrial magmatic rock that the resemblance is uncanny, yet the results, excellent as they are in many cases, have commonly been fitted somewhat uncomfortably into the framework of Gilbert's impact theory, the volcanic rocks being still popularly explained as secondary, impact-induced magmas. Any glass present in the lunar samples tends to be considered as of necessity of impactite origin, yet only the minutest fraction of terrestrial glasses are of this origin, and most glasses *associated with basalts* are volcanic! The age-dating results from the Moon are unexplained: that the 'soil' should be older than the enclosed 'basalt fragments' is extraordinary. Attempts to introduce a mysterious impact-explosion scattered exotic, older rock fraction KREEP, enriched in potash, rare Earth elements, and phosphorus, and emanating from rayed craters on the lunar highlands, to explain the anomaly, are suspect. The microscopic areas of KREEP 'recognised' by some investigators in lunar samples look suspiciously like the interstitial mesostasis of normal doleritic or gabbroic rocks on Earth, and are probably not introduced at all, but part of the rock as originally formed. The lunar soil is probably a primary volcanic breccia modified by impact: such a rock could well give different radiometric ages for the various 'elements' that make up the whole rock. What seems called for is careful study of comparable terrestrial material.

We know now, despite doubts about the anomalies mentioned above, that the immense-scale cratering phenomena that produced the maria occurred at least more than 3,000 million years ago, and probably much earlier. Of critical importance would be an Apollo mission to Tycho or Copernicus, which would show whether even these last mega-events of the lunar surface come within our familiar geological time scale. One of the problems emerging from Apollo studies is that the greater part of the mega-cratering including giant maria formation has to have taken place very rapidly and very soon after accretion, and a rigid crust must have been established by then. Endogenous mega-cratering does not have the same requirement, not being necessarily essentially explosive in nature. We should consider seriously the fluidisation cratering analogies of Mills, acting very early on a Moon, possibly without a thick rigid crust established, during a primitive degassing episode.

Urey and his followers have always favoured a cold lunar interior, and he has only moved so far, in considering the Apollo evidence, to entertain an inhibition of internal melting because of the small size of the Moon, coupled with dermal melting of an 'achondrite' layer. Many other scientists, however, now accept deep-seated internal heating of the Moon, though being uncertain of its cause. The geophysicists are divided as to whether the evidence of artificial and natural lunar seisms indicates a heated interior, as Latham has suggested, or a cold interior, as Sonnett concludes; but there is accruing evidence for a still active (albeit weakly so) Moon in tidally triggered transient events, and detection of actual gas releases: and it is becoming clear that the 'lava' that filled the maria is a primary magma.

The Moon should theoretically have been hit by meteorites repeatedly, and having no atmosphere, these should have arrived with unabated cosmic velocity, to explode on impact, forming many

312

craters, large and small, and vaporising. However, we know little of the flux of meteorites beyond a million years ago, and as yet we know little of the effect of high velocity impacts on the lunar surface —the physical scar produced on the surface by the artificial impact of *Intrepid* at 134 m/s has not, so far, been described in detail. Despite the correctness of the theoretical basis, there is much evidence that the giant lunar maria, immense circular to irregular depressions up to 400 km across, are not astroblemes, but are reflections of a primitive eruptivity—possibly analogous with ocean-basin eruptivity on Earth. No one doubts that small impact-explosion craters are likely to exist on the surface of the Moon, but even so, careful analysis of the structures that liberally pock the lunar surface, their control by tectonic grain and their distribution, does not unequivocally support an impact origin for the majority—there seems to be a good alternative case for an origin in degassing, in eruptions of predominantly gaseous material carrying up rock blocks, and fusing the bottoms of the craters by the agency of the heat of the rising gas.

The meteoritic and volcanic theories for the lunar cratering really hinge, however, on the origin of the giant craters. Again, direct studies of the rayed craters such as Tycho and the nature of the rays themselves, which from Ranger 7 and other satellite evidence seem to be nothing more than light areas produced by clustered, minute craters, will be critical to a final resolution of the age-old controversy.

The Mariner 6, 7, and 9 missions to Mars have done nothing to strengthen the meteoritic or 'ballistic' theory. Areas of the surface are covered by craters in identical or 'fraternal' (apparently contemporaneous, but of different size) crater pairs of great size. The abundance of crater pairs seems quite inexplicable in terms of ballistic theory, for one simply cannot envisage repeated incidences on this scale of equal sized pairs or large-small pairs of meteoroids of immense dimensions 'hugging one another' in space, so as to produce giant craters that are actually tangentially in contact. Could meteoroids really travel in this way in space? Atmospheric break up is not relevant to this anomaly. Most scientists agree that Nix Olympica, a feature recognised by Mariner 9 in 1972, is a 500 mile diameter volcano.

Apollo probes have located small traces of meteoritic metal in the lunar regolith, and geochemists have adduced a 2 per cent meteoritic component (a conclusion not entirely substantiated), yet the actual visible meteoritic material on the surface is small in quantity, and

313

there is evidence that the micrometeorite flux has been less than expected. If the ballistic theory found some peripheral support in the discovery by Apollo 11 and 12 astronauts of small craters with glass spheres concentrated in the bottom of the bowl and fused areas in the walls, even the assumption that this glass is necessarily an impactite seems to have been made over-hastily. Fusion by hot ascending gas could just as well produce these phenomena. These glasses will be carefully studied, and it will be interesting to see whether they contain significant traces of nickel-iron from the fused meteorite, or traces of the supposed exotic secondary impacting material, which must also have fused. If the results are contrary to this expectation, the uprise of hot volcanic gases, mostly dry, and not exactly like volcanic gases on Earth, up numerous fracture lines that transect the lunar surface multitudinously, must be seriously entertained. The glass spheroids, fused crater walls and ejected blocks could all be accommodated in such a theory, which could also possibly more easily accommodate the present age-dating anomalies.

There is evidence in the nature of the rilles (Apollo 15), and in a recent detection of vapour discharges that there is emission continuing even now from the lunar surface fractures. The mare filling could be a primary volcanic breccia emplaced largely by vertical feeding (with or without significant lateral spreading: the consistency of engine oil suggested for these magmas would allow lateral flows without significant physiographic expression). It could have been reworked by later gas eruptions bringing up blocks of more recent crystallisation, which became churned in with the original breccia to form the present 'soil', and also modified to a limited extent by impact. Impact agencies of fusion and brecciation have probably been over-emphasised in lunar studies; and pyroclasis correspondingly neglected.

No one now doubts the reality of lunar volcanism: the problem is 'was it secondary and consequent on mare formation by extraterrestrial agency, or was it primary?' The lunar basalts are so like the primary terrestrial basalts and the basaltic achondrite meteorite material (always accepted as primary), that it is difficult to see the need for a special ballistic explanation for the Moon, except to explain the mega-cratering. Amstutz has lately referred to 'text-books organised more like bibles than scientific books'—and raised spirited counter-criticism from the establishment! It is true that science *is* inhibited by doctrinaire establishments, and any open-minded

314

scientist will know that it is so. In no modern field of science is this so true as in lunar science. The evidence against the ballistic theory still remains largely neglected. The tidal periodicity of the lunar transient events, the seismic periodicity that matches it, the whole nature of the lunar surface when studied in detail, the magmatic nature of the lunar samples, the fact that the plagioclasic basalt and anorthosites of the highlands can only reasonably stem from mantle melting on a global scale, all this is indicating that the Moon possesses internal eruptivity involving heating and primary magmatism, pyroclasis, and gas emission—*sui generis*, but not altogether unlike the processes within other bodies of the solar system.

The largest known meteorite weighs only 60 tons, we know nothing of the ancient meteorite flux, and meteorite cosmic ray exposure ages do nothing to support lunar impact cratering (the parent body seems to have broken up much later than most of the lunar cratering!). Though we can, like Ringwood, invoke an initial post-accretion sweeping up of planetesmals to produce the craters, this is to invoke an earlier generation of meteoroids. The alternative invocation of a primitive eruptive phase for the small planet and satellite, Mars and the Moon, involving much out-gassing, should not, as yet, be discounted. The idea is attractive because, as already noted, it could be coupled with the enigmatic disruption of the even smaller asteroids. The Earth may either have suffered a similar primitive out-gassing and surface cratering, or being a much larger globe, may have developed quite differently. Phobos may show us just what a meteorite parent body looked like after some disruption—whether a captured asteroid or not.

And so, in this book dealing with meteorites, the author must depart from popular belief and state that, though many believe the giant lunar craters and maria to be impact explosion scars, he cannot see in them a certain extension of the meteoritic phenomena known to man, even though some of the smaller lunar and Martian craters are likely to be impact scars, and the giant craters may yet, against his conclusions, prove also to be such scars.

Bibliography

General

1 Mason, B. *Meteorites*. John Wiley, New York, 272 pp, 1962.
2 Krinov, E. L. *Principles of Meteorites*. Pergamon, New York, 535 pp, 1960.
3 Fedynsky, V. V. *Meteors*. Foreign Languages Publishing House, Moscow, 126 pp, 1959.
4 Nininger, H. H. *Out of the Sky*. Dover Publications, New York, 336 pp, 1959.
5 Heide, F. *Meteorites*. University of Chicago Press, 144 pp, 1957.
6 Anders, E. 'Meteorite Ages', in *The Solar System—IV, Moon, Meteorites and Comets*, eds. Kuiper, G. P. and Middlehurst, B. University of Chicago Press, pp 402–495, 1963.
7 Wood, J. A. 'Physics and Chemistry of Meteorites', *ibid*, pp 337–401, 1963.
8 Krinov, E. L. 'The Tunguska and Sikot-Alin Meteorites', *ibid*, pp 208–234, 1963.
9 Krinov, E. L. 'Meteorite Craters on the Earth's Surface', *ibid*, pp 183–207, 1963.
10 Perry, S. H. 'The metallography of meteoritic iron'. *Bull. U.S. Natl. Museum*, **184**, 206, 1944.
11 Axon, H. J. 'The metallurgy of meteorites'. *Progress in Materials Science*, **13**, 184–228, 1968.
12 Ringwood, A. E. 'Chemical and genetic relationships among meteorites'. *Geochimica et Cosmochimica Acta*, **24**, 159–197, 1961.
13 Ringwood, A. E. 'Chemical evolution of the terrestrial planets', *ibid*, **30**, 41–104, 1966.
14 Mason, B. 'Meteorites'. *American Scientist*, **55**, 429–455, 1967.
15 Baker, G. 'Tektites'. *Mem. Natl. Museum Victoria*, **23**, 313, 1959.

16 O'Keefe, J. A. (ed.) *Tektites*. Univ. of Chicago Press, 228 pp, 1963.

17 Fletcher, L. *Introduction to the Study of Meteorites*. British Museum, HMSO, London, 1904.

18 Tschermak, G. 'The microscopic properties of meteorites'. Transl. J. A. & E. M. Wood, *Smithsonian Contrib. in Astrophysics*, **4**, 138–239, 1964 (original publication, 1885).

Specific by Chapter

(References given previously in general list or earlier chapter lists are *not* repeated.)

Chapter 2

Bøggyld, O. B. 'Iglukunguaq nickeliferous pyrrhotite'. *Meddelelsser om Grønland*, **32**, 1905.

Chapter 3

Cepelecha, Z. 'Multiple fall of the Pribram meteorites photographed'. *Bull. Astron. Inst. Czechoslovakia*, **12**, 21–47, 1961.

Porter, J. G. 'The statistics of comet orbits', in *The Solar System—IV, Moon, Meteorites, and Comets*, eds Kuiper, G. P. and Middlehurst, B., University of Chicago Press, pp 550–572, 1963.

Roth, G. D. *The System of the Minor Planets*. Faber and Faber, 128 pp, 1962.

Schmidt, O. J. *A Theory of the Earth's Origin*. Foreign Languages Publishing House, Moscow, 139 pp, 1958.

Whipple, F. L. 'On the Structure of the Cometary Nucleus', in *The Solar System—IV, Moon, Meteorites and Comets*, eds Kuiper, G. P. and Middlehurst, B., University of Chicago Press, pp 639–664, 1963.

Chapter 4

Bunch, T. E. and Cassidy, W. A. 'Impact induced deformation in the Campo del Cielo meteorite', in *Shock Metamorphism in Natural Materials* (French, B. M. and Short, N. M. eds). Mono Book Corpn, Baltimore, pp 610–612, 1968.

Cassidy, W. A., Villar, L. M., Bunch, T. E., Kohman, T. P., and Milton, D. J. 'Meteorite craters of Campo del Cielo, Argentina'. *Science*, **149**, 1055–1064, 1965.

Madigan, C. T. 'The Boxhole crater and Huckitta meteorite'. *Trans. Roy. Soc. S. Australia*, **61**, 187–190, 1937.

McCall, G. J. H. 'New material from and a reconsideration of the Dalgaranga meteorite and crater, Western Australia'. *Min. Mag.*, **35**, 476–487, 1965.

McCall, G. J. H. and Jeffery, P. M. 'The Wiluna meteorite fall, Western Australia'. *Min. Mag.*, **37**, 880–887, 1970.

Milton, D. J. and Michel, F. C. 'Structure of a ray crater at Henbury, Northern Territory, Australia'. U.S.G.S. Prof. Paper, 525c, 1965.

Monod, T. H. and Pourquoie, A. 'Le cratère d'Aouellouel, Adrar, Sahara Occidentale'. *Bull. Inst. Franc. Afrique Noire*, **13**, 293–311, 1951.

Nininger, H. H. 'The Odessa meteorite crater'. *Sky and Telescope*, **III**, No. 11, 1939.

Nininger, H. H. 'The excavation of a meteorite crater near Haviland, Kiowa Co., Kansas'. *Proc. Colorado Mus. Nat. Hist.*, **12**, 1933.

Nininger, H. H. 'Arizona's meteorite crater'. American Meteorite Museum Publication, Sedona, Arizona, 232 pp, 1936.

Nininger, H. H. and Huss, G. I. 'The unique meteorite crater at Dalgaranga, Western Australia'. *Min. Mag.*, **32**, 619–639, 1960.

Spencer, L. J. 'Hoba, South West Africa—the World's largest known meteorite'. *Min. Mag.*, **23**, 1–18, 1932.

Spencer, L. J. 'Meteoritic iron and silica glass from the meteoritic craters of Henbury, Central Australia and Wabar, Arabia'. *Min. Mag.*, **23**, 387–404, 1933.

Chapter 5

Beck, C. and LaPaz, L. 'The nortonite fall and its mineralogy'. *American Mineralogist*, **36**, 45–49, 1951.

Buchwald, V. F. 'A new Cape York meteorite discovered'. *Geochimica et Cosmochimica Acta*, **28**, 125–126, 1964.

Clarke, R. S., Jaresowich, E., Mason, B., Nelen, J., Gomez, M., and Hyde, J. R. *The Allende, Mexico, Meteorite Shower*. Smithsonian contributions to the Earth Sciences, No 5, 53 pp, 1970.

Hey, M. H. *Catalogue of Meteorites*. British Museum (Nat Hist), HMSO, 637 pp, 1966.

McCall, G. J. H. and de Laeter, J. R. *Catalogue of Western Australian Meteorite Collections*. Spec Publ No 3, Western Australian Museum, 138 pp, 1965.

McCall, G. J. H. First supplement to the above, 18 pp, 1968.

McKellar, J. B., Meadows, A. J., and Sylvester-Bradley, P. C. 'The Barwell meteorite'. *Trans. Leicester Literary and Philosophical Society*, **60**, 22–28, 1966.

Miles, H. G. and Meadows, A. J. 'The Barwell meteorite'. *Nature, Lond.*, **210**, 983, 1966.

Chapter 6

Brezina, A. 'The arrangement of collections of meteorites'. *Proc Am. Phil. Soc.*, **43**, 211–247, 1904.

McCall, G. J. H. 'The Bencubbin Meteorite—further details including microscopic character of host material and two chondritic enclaves'. *Min. Mag.*, **36**, 726–739, 1968.

Mason, B. and Wiik, H. B. 'Amphoterites and meteorites of similar composition'. *Geochimica et Cosmochimica Acta*, **28**, 533–538, 1964.

Prior, G. T. 'The classification of meteorites'. *Min. Mag.*, **19**, 51–63, 1920.

Rose, G. 'Beschreibung und Entheilung der Meteoriten auf Grund der Sammlung im mineralogischen Museum zu Berlin'. *Physik. Abhandl. Akad. Wiss. Berlin*, 23–161, 1863.

Tschermak, G. 'Beitrag zur Classification der Meteoriten'. *Sitzber. Akad. Wiss. Wien*, Math-naturw. Kl., Abt. 1, **88**, 347–371, 1883.

Van Schmus, W. R. and Wood, J. A. 'A chemical-petrologic classification for chondritic meteorites'. *Geochimica et Cosmochimica Acta*, **31**, 747-765, 1967.

Chapter 7

LaPaz, L. 'The effects of meteorites upon the Earth, including its inhabitants, atmosphere, satellites'. *Adv. Geophys.*, **4**, 217–350, 1958.

Lovering, J. F. 'Frequency of meteorite falls throughout the ages'. *Nature, Lond.*, **183**, 1664–1665, 1959.

Chapter 8

Anderson, C. A., Keil, K., and Mason, B. 'Silicon oxynitride—a meteorite mineral'. *Science*, **146**, 256–257, 1964.

Bannister, F. A. 'Osbornite, meteoritic titanium nitride'. *Min. Mag.*, **26**, 36–44, 1941.

Dawson, K. R., Maxwell, J. A., and Parsons, D. E. 'Description of a meteorite which fell near Abee, Alberta, Canada'. *Geochimica et Cosmochimica Acta*, **21**, 127–144, 1960.

Du Fresne, E. R. and Roy, S. K. 'A new phosphide mineral from the Springwater pallasite'. *Geochimica et Cosmochimica Acta*, **24,** 198–205, 1961.

Mason, B. 'Minerals of meteorites', in *Researches on meteorites*, ed. Clarke, C. B., John Wiley, New York, pp 145–163, 1962.

Mason, B. 'Extraterrestrial mineralogy'. *Am. Mineral.*, **52,** 307–327, 1967.

Ramdohr, P. 'The opaque minerals in stony meteorites'. *J. Geophys. Res.*, **68,** 201–236, 1963.

Ringwood, A. E. 'Silicon in the metal phase of enstatite chondrites'. *Geochimica et Cosmochimica Acta*, **25,** 1–13, 1961.

Sztrokay, K. I. 'Uber einige Meteoritenmineralien des Kohlenwasserstoffhaltigen Chondrites von Kaba, Ungarn'. *Neus Jahrb. Mineral. Abhandl.*, **94,** 1284–1294, 1960.

Chapter 9

Goldstein, J. I. and Short, J. M. *The Iron Meteorites, Their Thermal Histories and Parent Bodies.* Publication of Goddard Space Flight Center, Maryland, 69 pp, 1967.

Goldstein, J. L. and Ogilvy, R. E. 'The growth of Widmastätten Pattern in metallic meteorites'. *Geochimica et Cosmochimica Acta*, **29,** 893–920, 1965.

Henderson, E. P. 'Hexahedrites'. *Smithsonian miscellaneous collection*, **148,** No 5, 1–41, 1965.

Lovering, J. F. 'Differentiation in the iron-nickel core of a parent meteorite body'. *Geochimica et Cosmochimica Acta*, **12,** 238–252, 1957.

Lovering, J. F. 'Pressure and temperature within a typical meteorite parent body'. *Ibid*, **12,** 253–261, 1957.

Uhlig, H. H. 'Contribution of metallurgy to the study of meteorites Part 1—Structure of metallic meteorites, their composition and the effect of pressure'. *Geochimica et Cosmochimica Acta*, **6,** 282–301, 1954. Part II—'Significance of Neumann Bands in meteorites'. *Ibid*, **7,** 34–42, 1955.

Chapter 10

Cleverly, W. H. 'Further recoveries of two impact-fragmented Western Australian Meteorites, Haig and Mt. Egerton'. *J. Roy. Soc. W. Aust.*, **51,** 76–88, 1968.

Coulson, A. L. 'The Patwar meteoric shower'. *Records Geol. Surv. India*, **69**, 439–457, 1936.

Hess, H. H. and Henderson, E. P. 'The Moore county meteorite—a further study with comment on its primordial environment'. *Am. Mineral.*, **34**, 494–507, 1949.

Lovering, J. F. 'Evolution of the meteorites—evidence for the co-existence of chondritic, achondritic and iron meteorites in a typical parent meteorite body', in *Researches on Meteorites*, ed Moore, C. B., John Wiley, New York, pp 179–197, 1962.

McCall, G. J. H. 'Evidence for the unity of provenance of true meteorites and against derivation of certain aerolite groups from the Moon'. *Transactions of the lunar geological field conference*, Bend, Oregon, Sponsored by the University of Oregon and New York Academy of Sciences, 43–48, 1963.

McCall, G. J. H. 'A meteorite of unique type from Western Australia —The Mt. Egerton Stony Iron'. *Min. Mag.*, **35**, 241–249, 1965.

McCall, G. J. H. 'The petrology of the Mt. Padbury mesosiderite and its achondrite enclaves'. *Min. Mag.*, **35**, 1029–1060, 1966.

Mason, B. *The Pallasites*. American Museum Novitates, No 2163, 19 pp, 1963.

Mason, B. 'The Woodbine meteorite'. *Min. Mag.*, **35**, 120–126, 1967.

Chapter 11

Keil, K. 'Mineralogical and chemical relationships among enstatite chondrites'. *J. Geophys. Res.*, **73**, 6945–6976, 1968.

McCall, G. J. H. and Cleverly, W. H. 'New meteorite finds including two ureilites, from the Nullarbor Plain, Western Australia'. *Min. Mag.*, **36**, 619–716, 1968.

Mason, B. 'The Bununa meteorite, and a discussion of the pyroxene-plagioclase achondrites'. *Geochimica et Cosmochimica Acta*, **31**, 107–115, 1967.

Mason, B. *The Hypersthene Achondrites*. American Museum Novitates, 2155, 13 pp, 1963.

Mason, B. 'The carbonaceous chondrites'. *Space Science Reviews*, **1**, 621–646, 1962–1963.

Mason, B. 'The amphoterites and meteorites of similar composition'. *Geochimica et Cosmochimica Acta*, **28**, 533–538, 1964.

Mason, B. *The Chemical Composition of the Olivine-Bronzite and Olivine-Hypersthene Chondrites*. American Museum Novitates, 2223, 38 pp, 1965.

Mason, B. 'The enstatite chondrites'. *Geochimica et Cosmochimica Acta*, **30,** 23–29, 1966.

Mason, B. *The Classification of Chondritic Meteorites.* American Museum Novitates, 2085, 20 pp, 1962.

Mason, B. 'Olivine composition in chondrites'. *Geochimica et Cosmochimica Acta*, **27,** 1011–1023, 1963 and **31,** 1100–1103, 1967.

Ringwood, A. E. 'Silicon in the metal phase of enstatite chondrites and some geochemical implications'. *Geochimica et Cosmochimica Acta*, **25,** 1–13, 1961.

Vdovykin, G. P. 'Ureilites'. *Space Science Reviews*, **10,** 483–510, 1970.

Wasson, J. T. and Wai, C. M. 'Composition of metal, schreibersite and perryite of enstatite achondrites and the origin of enstatite chondrites and achondrites'. *Geochimica et Cosmochimca Acta*, **34,** 169–184, 1970.

Chapter 12

Goldberg, E., Uchiyama, A., and Brown, H. 'The distribution of nickel, cobalt, gallium, palladium and gold in iron meteorites'. *Geochimica et Cosmochimica Acta*, **2,** 1–25, 1951.

Goldschmidt, V. M. *Geochemische Verteilungsgestze der Elemente*, No 11, Norske Videnskaps. Skrifter, Mat-Naturv. Klasse, No 4, 148 pp, 1938.

Keil, K. and Fredriksson, K. 'The iron, magnesium and calcium distribution in co-existing olivines and rhombic pyroxenes of meteorites'. *J. Geophys. Res.*, **69,** 3487–3517, 1964.

Mason, B. 'Geochemistry and meteorites'. *Geochimica et Cosmochimca Acta*, **30,** 365–374, 1966.

Nordenskjold, A. E. 'Trenne markeliga eldmeteorer, sedda i Sverige under åren 1876 och 1877'. *Geol. Foreh. i Stockholm Forh.*, **4,** 45–61, 1878.

Prior, G. T. 'The meteoritic stones of Launton, Warbreccan, Cronstad, Daniel's Kuil, Khairpur and Soko-Banja'. *Mineral. Mag.*, **18,** 1–25, 1916.

Urey, H. C. and Craig, H. 'The composition of the stone meteorites and the origin of the meteorites'. *Geochimica et Cosmochimica Acta*, **4,** 36–82, 1953.

Wasson, J. T. 'The chemical classification of iron meteorites'. *Geochimica et Cosmochimica Acta*, **31,** 161–180, 1967; **31,** 2065–2093, 1967; **33,** 859–876, 1969; *Icarus*, **12,** 407–423, 1970.

323

Wiik, H. B. 'The chemical composition of some stony meteorites'. *Geochimica et Cosmochimica Acta*, **9**, 279–289, 1956.

Wiik, H. B. 'On regular discontinuities in the composition of meteorites'. *Commentationes Physico-Mathematicae*, **34**, 135–145, 1969.

Wiik, H. B. *On the Genetic Relationship between Meteorites.* Arizona State University, Tempe, Arizona, Contribution No 13, Center for meteorite studies, 14 pp, 1966.

Chapter 13

Binns, R. A. 'Chondritic inclusion of unique type in the Cumberland Falls meteorite', in *Meteorite Research* (ed P. M. Millman), Reidel, Dordrecht, Holland, pp 696–704, 1969.

Fredriksson, K. and Keil, K. 'Light-dark structure in the Pantar and Kapoeta stone meteorites'. *Geochimica et Cosmochimica Acta*, **27**, 717–740, 1963.

Mason, B. 'The Woodbine meteorite, with notes on silicates in iron meteorites'. *Min. Mag.*, **36**, 120–126, 1967.

Olsen, E. and Jaresowich, E. 'Chondrules: first occurrence in an iron meteorite'. *Science*, **174**, 583–585, 1971.

Wahl, W. 'The brecciated stony meteorites and meteorites containing foreign fragments'. *Geochimica et Cosmochimica Acta*, **2**, 91–117, 1952.

Wasson, J. T. 'Ni, Ga, Ge and Ir in the metal of iron meteorites with silicate inclusions'. *Ibid*, **34**, 957–964, 1970.

Chapter 14

Nagy, B. 'Carbonaceous meteorites'. *Endeavour*, **27**, 81–86, 1968.

Chapter 15

Fredriksson, K. 'Chondrules and the meteorite parent bodies'. *Trans. N.Y. Acad. Sci.*, Ser 2, **25**, 756–759, 1963.

Mason, B. 'Origin of chondrules and chondritic meteorites'. *Nature, Lond.*, **186**, 230–231, 1960.

Mueller, G. 'Interpretation of misconstructures in Carbonaceous Chondrites'. *Advances in Organic Geochemistry* (eds Colombo, U., and Hobson, G. D.), pp 119–140, Pergamon Press, Oxford, 1964.

Mueller, G. 'The properties and theory of genesis of the carbonaceous complex within the Cold Bokkeveld meteorite'. *Geochimica et Cosmochimica Acta*, **4**, 1–10, 1953.

Ringwood, A. E. 'Genesis of chondritic meteorites'. *Reviews of Geophysics*, **4**, 113–175, 1966.

Wood, J. A. 'Chondrites and the origin of the terrestrial planets'. *Nature, Lond.*, **194**, 127–130, 1962.

Wood, J. A. 'On the origin of chondrules and chondrites'. *Icarus*, **2**, 152–180, 1963.

Wood, J. A. 'Properties of chondrules', in *Origin of the Solar System*, eds Jastrow, R. and Cameron, A. G. W., Academic Press, New York and London, pp 147–154, 1963.

Chapter 16

Fredriksson, K. and Mason, B. 'The Shaw meteorite'. *Geochimica et Cosmochimica Acta*, **31**, 1705–1709, 1967.

McCall, G. J. H. and Cleverly, W. H. 'A review of meteorite finds on the Nullarbor Plain, Western Australia, including a description of thirteen new finds of stony meteorites'. *J. Roy. Soc. W. Aust.*, **53**, 69–80, 1970.

Chapter 17

Claus, G. and Nagy, B. 'A microbiological examination of some carbonaceous chondrites'. *Nature, Lond.*, **192**, 594–596, 1961.

Hayes, J. M. 'Organic constituents in meteorites'. *Geochimica et Cosmochimica Acta*, **31**, 1395–1440, 1967.

Meinschein, W. G. Quarterly report on NASA contract NASW 508 1 July 1964.

Nagy, B., Meinschein, W. G., and Hennessy, D. J. 'Mass spectrographic analysis of the Orgueil meteorite: evidence for biogenic hydrocarbons'. *Trans. N.Y. Acad. Sci.*, **93**, 25–35, 1961.

Chapters 18 and 19

Anders, E. 'Meteorites and the early history of the solar system', in *Origin of the Solar System*, eds Jastrow, R. and Cameron, A. G. W., Academic Press, New York and London, pp 95–142, 1963.

Anders, E. 'Origin age and composition of meteorites'. *Space science reviews*, 583–714, 1964.

Anders, E. 'Meteorites and the early solar system'. *Annual Review of astronomy and astrophysics*, **9**, 1–34, 1971.

Goles, G. G., Fish, R. A., and Anders, E. 'The record in the meteorites I. The former environment of stone meteorites as deduced from K40–Ar40 ages'. *Geochimica et Cosmochimica Acta*, **19**, 177–195, 1960.

Chapters 21 and 22

Adams, E. W. and Huffaker, R. M. 'Application of ablation analysis to stony meteorites and the tektite problem'. *Nature, Lond.*, **143**, 1249–1251, 1962.

Baker, G. 'The role of aerodynamic phenomena in the sculpturing and shaping of tektites'. *Am. J. Sci.*, **256**, 369–383, 1958.

Baker, G. 'Origin of tektites'. *Nature, Lond.*, **185**, 291–294, 1960.

Barnes, V. E. 'North American Tektites'. *Univ. of Texas Publ., Contrib. to Geology*, part 2, pp 477–582, 1939.

Barnes, V. E. 'Tektites'. *Scientific American*, **205**, 8, No 5, 58–65, 1961.

Chao, E. C. T., Adler, I., Dwornik, E. J., and Littler, J. 'Metallic spherules in tektites from Isabella, Philippine Islands', *Science*, **135**, No 3498, 97–98, 1962.

Chapman, D. R. and Larson, H. K. *Lunar origin of Tektites*. National Aeronautics and Space Administration, Tech. Note D 1556, Washington, 66 pp, 1963.

Darwin, C. *Geological Observations on the Volcanic Islands and Parts of South America during the Voyage of H.M.S. Beagle*. Republication, Appleton & Company, N.Y., pp 44–45, 1891.

Lindaker, J. T. Nachtrage und Zusätze den böhmischen Topasen und chrysolithen. Sammlung physikalischer Aufsabe besonders die böhmischen Naturforscher heransgegeben von Johann Mayer. 2nd Band, 272, Dresden, 1792.

Kuiper, G. P. 'Satellites, comets and interplanetary material'. *Proc. Natl. Acad. Sci. Washington*, **40**, 1153–1158, 1953; and 1096–1112, 1954.

McCall, G. J. H. 'Tektites—the conflict of evidence'. *Planetarium*, **1**, 175–179, 1968.

McColl, D. H. and Williams, G. E. 'Australite distribution pattern in Southern Central Australia'. *Nature, Lond.*, **226**, 154–155, 1970.

Schnetzler, C. C., Philpotts, J. A., and Thomas, H. H. 'Rare earth and barium abundances in Ivory Coast tektites and rocks from the Bosumtwi crater, Ghana'. *Geochimica et Cosmochimica Acta*, **31**, 1987–1993, 1967.

Taylor, S. R. 'Distillation of alkali elements during the formation of australite flanges'. *Nature, Lond.*, **189**, 630–633, 1961.

Taylor, S. R. 'Significance of lunar data for the origin of tektites'. Abstract volume, Section 3, *43rd ANZAAS Congress*, Brisbane, pp 3–5, 1971.

Chapters 20 and 23

Amstutz, G. C. 'Impact, crypto-explosion or diapiric movements'. *Trans. Kansas Acad. Sci.*, **67**, 343–359, 1964.

Beals, C. S. 'A probable meteorite crater of Precambrian age at Holleford, Ontario'. *Publ. Dominion Observatory, Ottawa*, **24**, No 6, 117–142, 1960.

Beals, C. S., Innes, M. J. S., and Rottenberg, J. A. 'The search for fossil meteorite craters'. *Current Sci.*, **29**, 205–249, 1960.

Bucher, W. H. 'Cryptovolcanic structures in the United States'. *Records 16th International Geol. Congress*, **2**, 891–902, 1933.

Bucher, W. H. 'Cryptoexplosion structures from within or without the Earth (astroblemes or geoblemes)'. *Am. J. Sci.*, **261**, 597–649, 1963.

Bucher, W. H. 'Are cryptovolcanic structures due to meteoric impact?'. *Nature, Lond.*, **197**, 1241, 1963.

Bunch, T. E., Cohen, A. J., and Dence, M. R. 'Shock induced structural disorder in plagioclase and quartz'. *Contrib. Dominion Obs. Ottawa*, **8**, 509–518, 1968.

Bunch, T. E. and Short, N. M. 'A world wide inventory of features characteristic of rocks associated with presumed meteorite impact craters', in *Shock Metamorphism of Natural Material*. Eds French, B. M. and Short, N. M. Mono. Book Corpn., Baltimore, 255–266, 1968.

Cochran, W. 'Lunar Science Conference, Apollo 12'. *Geotimes*, **16** (2), 24–28, 1971.

Cook, P. J. 'The Gosses bluff cryptoexplosion structure'. *J. Geol.*, **76**, 123–139, 1968.

Currie, K. L. 'Rim structure of the New Quebec crater'. *Nature, Lond.*, **201**, 385, 1964.

Currie, K. L. 'Analogues of lunar craters on the Canadian Shield'. *Ann. N.Y. Acad. Sci.*, **123**, Art 2, 195–940, 1965.

Currie, K. L. and Dence, M. R. 'Rock deformation on the rim of the New Quebec crater'. *Nature, Lond.*, **198**, 80, 1963.

Currie, K. L. and Shafiqullah, N. 'Carbonatite and alkaline igneous rocks in the Brent crater, Ontario'. *Nature, Lond.*, **215**, 725–726, 1967.

Currie, K. L. and Shafiqullah, N. 'Geochemistry of some large Canadian craters'. *Nature, Lond.*, **218**, 457–459, 1968.

Dence, M. R., 1965. 'The extraterrestrial origin of Canadian craters'. *Ann. N.Y. Acad. Sci.*, **123**, Art 2, 941–969, 1965.

327

Dence, M. R. 'Shock zoning at Canadian craters, petrography and structural implications'. *Contrib. Dominion Obs., Ottawa,* **8,** No 26, 169–184, 1968.

Dence, M. R., Innes, M. J. S., and Robinson, P. B. 'Recent geological and geophysical studies of Canadian craters'. *Ibid,* **8,** No 25, 339–362, 1968.

Dietz, R. S. 'Shatter Cones in crypto-explosion structures (meteorite impact?)'. *J. Geol.,* **67,** 496–505, 1959.

Dietz, R. S. 'The Vredefort Ring—meteorite impact structure?'. *J. Geol.,* **69,** 499–516, 1961.

Dietz, R. S. 'Crypto-explosion structures—a discussion'. *Am. J. Sci.,* **261,** 650–664, 1963.

Dietz, R. S. 'Sudbury structure as an astrobleme'. *J. Geol.,* **72,** 412–434, 1964.

French, B. M. 'Possible relations between meteorite impact and igneous petrogenesis, as indicated by the Sudbury structure, Ontario'. *Bull. Volcan.,* **34,** 466–517, 1971.

Guppy, D. J., Brett, R., and Milton, D. J. 'Liverpool and Strangways craters, Northern Territory—two structures of probable impact origin'. *J. Geophys. Res.,* **76,** 5387–5393, 1971.

Hager, D. 'Crater Mound (Meteor Crater) Arizona, a geologic feature', *Am. Assoc. Petroleum Geologists, Bull.,* **37,** 821–857, 1953.

Hargraves, R. B. 'Shatter cones in the rocks of the Vredefort Ring'. *Geol. Soc. S. Afr., Trans.,* **64,** 147–153, 1961.

Innes, M. J. S., Pearson, W. J., and Geuer, J. W. 'The Deep Bay crater'. *Publ. Dominion Observatory, Ottawa,* **31,** 19–52, 1964.

Jensen, V. 'Early Precambrian impact structure and associated hyalomylonites near Agto, W. Greenland'. *Nature, Lond.,* **233,** 188–190, 1971.

Junner, N. R. 'The Geology of the Bosumtwi caldera, and surrounding county'. *Gold Coast Geol. Surv. Bull.,* **8,** 38 pp, 1937.

Karpoff, R. 'The meteorite crater of Talemzane, in Southern Algeria'. *Meteoritics,* **1,** 31–38, 1953.

Kulik, L. A. 'The question of the meteorite of June 30th, 1908 in Central Siberia'. *Pop. Astron.,* **45,** 561–562, 1937.

Lafond, E. C. and Dietz, R. S. 'Lonar Crater, India, a meteorite crater?'. *Meteoritics,* **2,** 111–116, 1964.

McCall, G. J. H. 'Are crypto-volcanic structures due to meteoritic impact?'. *Nature, Lond.,* **201,** 251–254, 1964.

McCall, G. J. H. 'The caldera analogy in selenology'. *Ann. N.Y.*

Acad. Sci., **123,** Art 2, 843–875, 1965.

McCall, G. J. H. 'The concept of volcano-tectonic undation in selenology'. *Adv. in Space Sci.*, Academic Press N.Y., **8,** 1–64, 1965.

McCall, G. J. H. 'An independent assessment of the Ranger 7–9 results'. *Trans. Lunar Geol. Field Conference*, Bend., Oregon, 33–42, 1966.

McCall, G. J. H. 'Possible meteorite craters—Wolf Creek Australia and analogs'. *Ann. N.Y. Acad. Sci.*, **123,** Art 2, 970–988, 1965.

McCall, G. J. H. 'Implications of the Mariner IV, Photography of Mars'. *Nature, Lond.*, **211,** 1384–1385, 1966.

McCall, G. J. H. 'Proof of volcano-tectonic origin of mare terrain on the Moon?'. *Nature, Lond.*, **223,** 275–276, 1969.

McCall, G. J. H. 'The lunar controversy'. *J. Brit. Astron. Ass.*, **80,** 19–29, 100–106, 190–199, 263–269, 358–360, and discussion **81,** 148–149, 1970–1971.

McCall, G. J. H. 'The surface of Mars'. *Astronomy Today*, **3,** 11–15, 1970.

McCall, G. J. H. (1970b). 'Lunar Rilles—a possible terrestrial analogue'. *Nature, Lond.*, **225,** 714–716, 1970.

Manton, W. I. 'Orientation and origin of shatter cones in the Vredefort Ring'. *Ann. N.Y. Acad. Sci.*, **123,** Art 2, 1017–1049, 1965.

Millman, P. M., Liberty, B. A., Clark, J. F., Willmore, P. C., and Innes, M. J. S. 'The Brent crater'. *Ibid*, **24,** No 1, 1–43, 1960.

Mills, A. A. 'Fluidisation phenomena and possible implications for the origin of lunar craters'. *Nature, Lond.*, **224,** 863–866, 1969.

Milton, D. J. and Naeser, C. W. 'Evidence for an impact origin for the Pretoria Salt Pan, South Africa'. *Nature, Phys. Science*, **229,** 211–212, 1971.

Rohleder, H. P. T. 'The Steinheim Basin and Pretoria Salt Pan— volcanic or meteoritic origin'. *Geol. Mag.*, **70,** 489–498, 1933.

Shoemaker, E. M. 'Impact mechanics at Meteor crater, Arizona', in *The Solar System—IV, Moon, Meteorites, Comets* (eds Kuiper, G. P. and Middlehurst, B.), pp 301–336, 1963.

Shoemaker, E. M. and Chao, E. C. T. 'New evidence for the impact origin of the Ries Basin, Bavaria, Germany'. *J. Geophys. Res.*, **66,** 3371–3378, 1961.

Snyder, F. G. and Gerdemann, P. E. 'Explosive igneous activity along an Illinois–Missouri–Kansas axis'. *Am. J. Sci.*, **263,** 465–493, 1965.

Index

331

93 Physci
JR